SAP S/4HANA Asset Management

Configure, Equip, and Manage your Enterprise

Rajesh Ojha
Chandan Mohan Jaiswal

Apress®

SAP S/4HANA Asset Management: Configure, Equip, and Manage your Enterprise

Rajesh Ojha
Atlanta, USA

Chandan Mohan Jaiswal
Singapore, Singapore

ISBN-13 (pbk): 978-1-4842-9869-5
https://doi.org/10.1007/978-1-4842-9870-1

ISBN-13 (electronic): 978-1-4842-9870-1

Managing Director, Apress Media LLC: Welmoed Spahr
Acquisitions Editor: Divya Modi
Development Editor: James Markham
Coordinating Editor: Divya Modi
Copyeditor: Kim Burton

Cover designed by eStudioCalamar

Cover image by Freepik (www.freepik.com)

Distributed to the book trade worldwide by Apress Media, LLC, 1 New York Plaza, New York, NY 10004, U.S.A. Phone 1-800-SPRINGER, fax (201) 348-4505, e-mail orders-ny@springer-sbm.com, or visit www.springeronline.com. Apress Media, LLC is a California LLC and the sole member (owner) is Springer Science + Business Media Finance Inc (SSBM Finance Inc). SSBM Finance Inc is a **Delaware** corporation.

For information on translations, please e-mail booktranslations@springernature.com; for reprint, paperback, or audio rights, please e-mail bookpermissions@springernature.com.

Apress titles may be purchased in bulk for academic, corporate, or promotional use. eBook versions and licenses are also available for most titles. For more information, reference our Print and eBook Bulk Sales web page at http://www.apress.com/bulk-sales.

Any source code or other supplementary material referenced by the author in this book is available to readers on GitHub (https://github.com/Apress). For more detailed information, please visit https://www.apress.com/gp/services/source-code.

Paper in this product is recyclable

Table of Contents

About the Authors

Rajesh Ojha is a Senior Advisory Consultant at IBM and a certified SAP S/4HANA EAM professional. He has more than 13 years of SAP consulting experience across various complex projects, roles, and responsibilities in oil and gas, construction, manufacturing, transportation, and chemicals and pharmaceuticals. His areas of expertise are in SAP asset management portfolio, enterprise digital transformation, and providing strategic consulting to business leaders on how to best leverage SAP's potential power in their organization.

Chandan Mohan Jaiswal is a manager in the technology consulting practice at EY, where he specializes in focusing on asset management and manufacturing solutions using the SAP S/4HANA Business Suite. He brings more than 11 years of experience in SAP consulting, supported by 6 years of industry experience in asset management, plant maintenance, and factory automation. Chandan has been involved in multiple implementations, transformations, and support projects, undertaking various roles and responsibilities for global clients across industries such as public transport, health science, manufacturing, and oil and gas.

About the Technical Reviewer

Prince Tyagi is a freelance technical reviewer specializing in absorbing a lot of data and articulating the most important points. He helps large technical organizations clearly communicate their message across multiple projects.

Tyagi has more than 13 years of experience in SAP ERP as a solution specialist in SAP PM, SAP TM, SAP iMRO, SAP CS, and SAP QM modules. He has worked on multiple E2E implementations, rollouts, enhancements, and support and upgrade projects in various domains, including retail, aviation, manufacturing, healthcare, chemical, defense, and the software industry.

He has extensive experience in interfaces, debugging, and integration with SAP GTS, SAP EHS modules, and other SAP modules.

CHAPTER 1

SAP S/4HANA Asset Management

Welcome to the world of SAP S/4HANA Asset Management, a comprehensive solution designed to streamline and enhance asset management processes within your organization. This chapter embarks on a journey through SAP S/4HANA Business Suite, focusing on its asset management capabilities. We delve into various topics that give you a deeper understanding of this powerful system.

The following are the key topics covered in this chapter.

- SAP S/4HANA Business Suite

- SAP S/4HANA Asset Management

- S/4HANA Asset Management key functionalities

- Positioning and integration of Asset Management with other SAP S/4HANA Business Applications

- Ways to provision SAP S/4HANA

- S/4HANA Asset Management user interface and user experience

© Rajesh Ojha and Chandan Mohan Jaiswal 2023
R. Ojha and C. M. Jaiswal, *SAP S/4HANA Asset Management*,
https://doi.org/10.1007/978-1-4842-9870-1_1

1.1. SAP S/4HANA Business Suite

SAP S/4HANA comprises business applications known as SAP Business Suite powered by the SAP HANA database. In contrast to the earlier Business Suite–SAP R/3 ECC–the S/4HANA business applications are built with a significantly simplified data model with a fewer number of tables, in-memory SAP HANA database, and new user interface technology called SAP Fiori, which are most important features of SAP S/4HANA Business Suite.

The SAP S/4HANA Business Suite (see Figure 1-1) comprises applications to seamlessly support various functions of an enterprise, such as engineering and change management, logistics, supply chain, asset management, finance, and customer service.

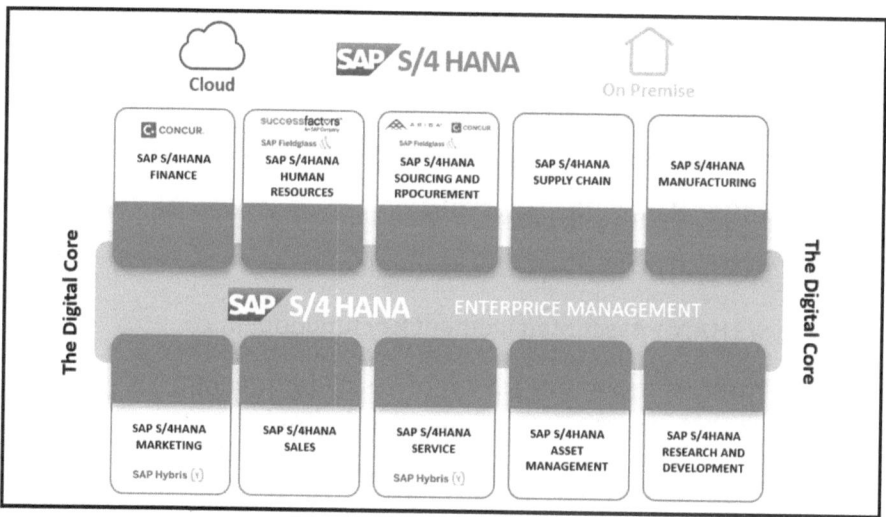

Figure 1-1. *S/4HANA offerings for enterprise management*

1.2. SAP S/4HANA Asset Management

One of the important goals of an enterprise is to optimize asset performance and maintain good health at the lowest possible cost. It is critical for an enterprise asset management application to deal with various aspects, as outlined in the following.

- Reliable, comprehensive, and time-phased master data management of assets

- Efficient ways of managing tasks, maintenance plans, and scheduling approaches to perform regular inspection, service, and repair

- Generating and monitoring real-time machine data

- Anticipating the performance and state of assets through machine learning and artificial intelligence

- Performing specific tasks on mobile devices

- Provision to network with various business partners through cloud-based platforms

- Comprehensive and integrated management of the entire lifespan of an asset

SAP S/4HANA Asset Management software is designed to encompass all the essential functionalities that can be utilized to address all the aspects as outlined. It offers the following independent solutions (also see Figure 1-2), which can be set up in combination for a complete and reliable asset management application.

- SAP S/4HANA Asset Management, public cloud

- SAP S/4HANA Asset Management, private cloud/on-premise

- SAP Intelligent Asset Management (SAP IAM)

- SAP S/4HANA Asset Management for resource scheduling

- SAP service and asset manager for mobile maintenance

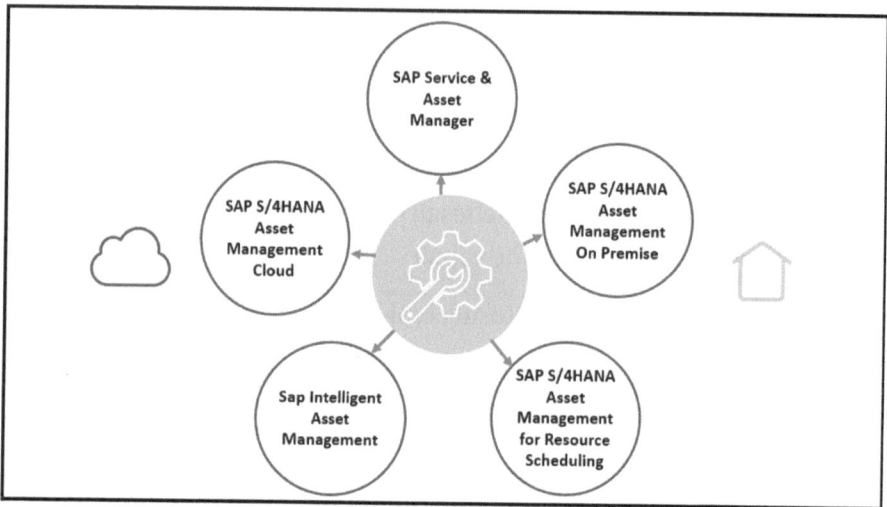

Figure 1-2. *S/4HANA Asset Management portfolio of solutions*

SAP IAM (Intelligent Asset Management SAP IAM) is a group of applications. These cloud-based solutions use the SAP Business Technology Platform. It has three parts: SAP Asset Intelligence Network, SAP Asset Performance Management, and SAP Service and Asset Manager. One big aim of SAP Intelligent Asset Management is to use technology and data science to predict when maintenance is needed. This helps to prevent unexpected problems and reduce the need for extensive maintenance.

1.3. S/4HANA Asset Management Key Functionalities

Like other business applications, the asset management software comprises four components concerning master data maintenance, business transactions, reporting and analytics, and access control (see Figure 1-3).

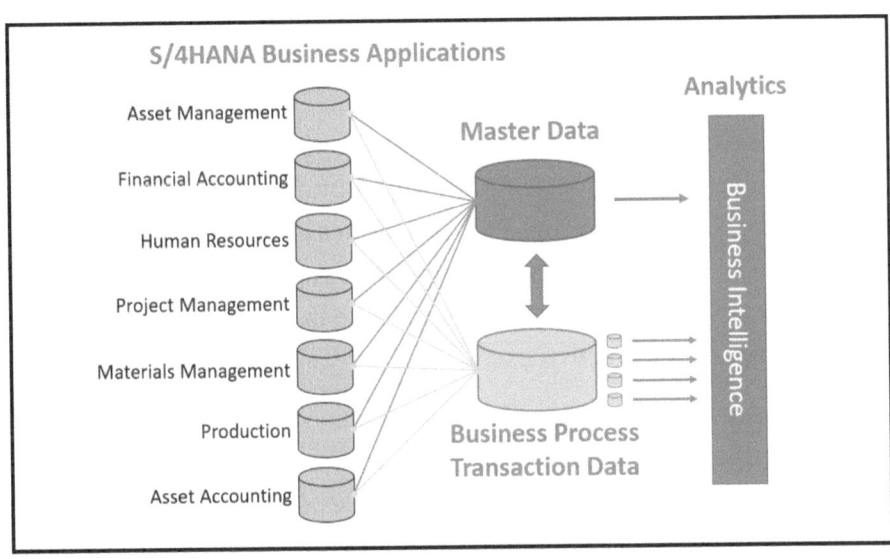

Figure 1-3. *Key functionalities in Asset Management application*

- **Master data maintenance**: Master data management is crucial in maintaining comprehensive records of technical assets and their associated characteristics. This encompasses asset details, maintenance planning and scheduling data, usage and performance values, bill of materials, resource capacity, and other relevant information. The primary objective of master data management is to guarantee the consistency, reliability, and adherence to data compliance standards.

5

- **Business process execution**: An enterprise asset management software aims to achieve a vital objective: enhance asset performance and ensure their well-being. It accomplishes this by supporting various business functions, including scheduled and unscheduled maintenance planning and scheduling, initiating, executing, and completing repair work, conducting quality inspections on technical objects, refurbishing assemblies, and performing other relevant tasks.

- **Reports and analytics**: Asset management reports and analytics serve the purpose of optimizing the tracking, monitoring, and analysis of assets, making these processes highly efficient. This capability enables businesses to derive maximum value from their asset management activities.

- **Access control**: The act of granting necessary authorizations to users enables access to specific resources. This serves a dual purpose of safeguarding sensitive data and ensuring that users can only access information and perform actions that are essential for their job responsibilities.

1.4. Positioning and Integration of Asset Management with Other SAP S/4HANA Business Applications

SAP S/4HANA Asset Management is developed with seamless integration with various other S/4 HANA Business applications, enabling automatic identification of pertinent business objects and automated posting of documents. The benefits of integrating asset management with other business applications include streamlined processes, real-time visibility, reduced downtime, and optimal spare parts management at the lowest possible cost.

Let's discuss a few of the important integrations.

Integration with S/4HANA Supply Chain applications focuses on materials management, including initiating material purchase requisitions, external service requisitions, and the management of inventory material from maintenance orders. Integration with other modules, such as Production Planning and Quality Management, enhances operational efficiency and inspection processes.

Integration with S/4HANA Financial Accounting and Controlling deals with the synchronization of the asset master from asset accounting and the equipment master from asset management to maintain accurate and consistent master data for reliable business transactions and reports on this data. It also involves the cost value flow in maintenance orders and the cost settlement of maintenance orders to cost objects in the controlling module.

Integration with the S/4HANA Project System enables linking maintenance orders with a project *work breakdown structure* (WBS). This integration offers precise cost tracking, especially during shutdowns, where maintenance costs can be allocated to specific project WBS, assisting in detailed cost analysis and reporting (see Figure 1-4).

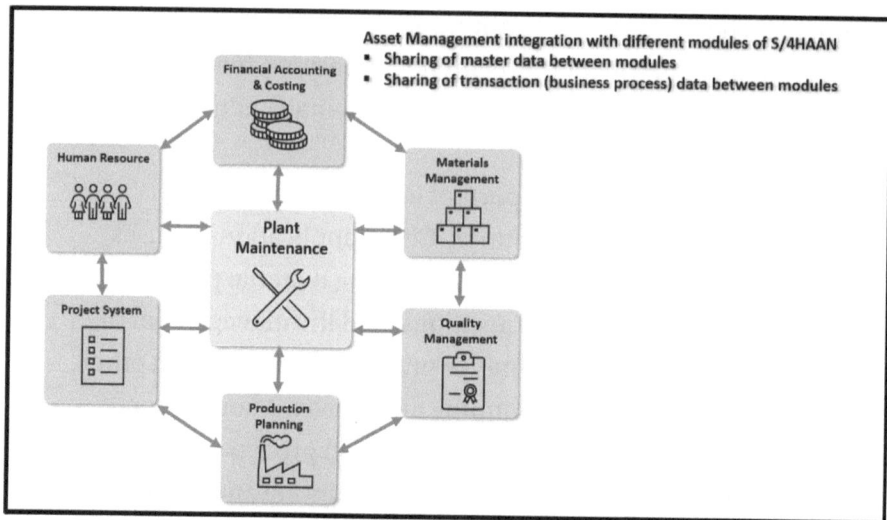

Figure 1-4. *Asset Management integration with other S/4HANA applications*

1.5. Ways to Provision SAP S/4HANA

The following are various ways to provision S/4HANA Asset Management and other business suite applications.

- **On-Premise**: In this widely used option, the customer purchases a license and installs the software in their own premise and data center. It is like having the SAP system in-house, and all the data and processes are managed and controlled internally by the company's IT team. This setup allows the company to have full control over its SAP system.

- **Cloud**: In this option, the customer rents the software for a defined time period. SAP provisions the software in its own premises and data center. This setup is like using a service or application online, and the company can access its SAP system from anywhere with an internet connection.

- **Hybrid**: This option involves a combination of customer-operated applications on-premise and SAP-operated applications in the cloud (see Figure 1-5). These two instances can be connected and seamlessly integrated with each other.

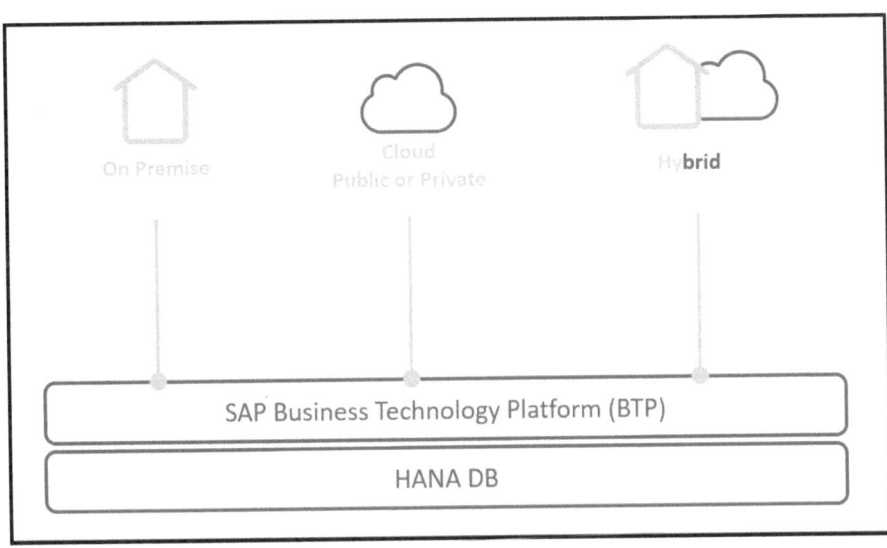

Figure 1-5. *Options to provision S/4HANA Business Suite*

1.5.1. Options Within Cloud Provisioning

The cloud provisioning model is available in two types: public cloud and private cloud software.

Public Cloud

In this offering, the software is typically hosted on a data center that serves multiple tenants or customers who share the server's resources. This setup is similar to an apartment building, where multiple tenants reside in the same physical infrastructure and utilize certain shared resources. However, each tenant has their own secure unit within the building, accessible only with their individual key. The maintenance of the building and apartment units is included in the rent paid by tenants and is handled by the cloud provider when necessary.

In the public cloud, each customer's data and applications are isolated from those of other customers. Despite this segregation, since they share certain resources and the cloud provider takes care of maintenance, the public cloud is often the most cost-effective and efficient solution.

Private Cloud

Private cloud software operates on a dedicated data center exclusively used by a single tenant or customer. The primary distinctions between private cloud and on-premise solutions lie in server maintenance responsibilities and software licensing.

With on-premises solutions, the customer purchases the server and assumes responsibility for its maintenance. In contrast, for private cloud setups, a third-party provider owns the server and takes care of its maintenance. Customers pay a subscription fee to access the server via the Internet and install the desired software. In certain cases, one cloud provider maintains the server, while another provider maintains the installed software.

For instance, in SAP S/4HANA Cloud, private edition, SAP is responsible for maintaining the business software. However, customers can host their software on a server within an SAP data center or in a data center operated by one of SAP's partners. In the latter case, the partner assumes the server maintenance duties. Private clouds offer greater flexibility and customization compared to public clouds. If a public cloud can be likened to an apartment, a private cloud is akin to a stand-alone house on its own land plot.

1.6. S/4HANA Asset Management User Interface and User Experience

A simple and easy-to-use user interface has become the driving factor behind the success of a software application. SAP continues to enhance user experience through continuous improvements and innovations in user interfaces. SAP S/4 HANA offers various user interface (UI) options.

1.6.1. SAP Fiori Apps

SAP Fiori apps is a new product that aims to achieve a modern, simplified, browser-based approach. SAP Fiori Launchpad (SAP FLP) is designed to be the entry point for all Fiori apps. SAP Fiori Launchpad supports SAPUI5 (HTML5) technology, SAP GUI for HTML, and Web Dynpro and runs in all the browsers available on the market. FLP is a role-based and personalized UI client that can be deployed on various platforms and device types. It is browser-based and consists of groups of tiles, which can be flexibly grouped based on tile catalogs. The availability of tile catalogs for each user is determined through role assignments.

1.6.2. SAP GUI for Windows

SAP GUI for Windows is client-level software of SAP and runs on the Microsoft Windows platform for all SAP applications (see Figure 1-6). It works like any browser and needs to be installed on local PCs within a company's network to remotely access the central SAP application server.

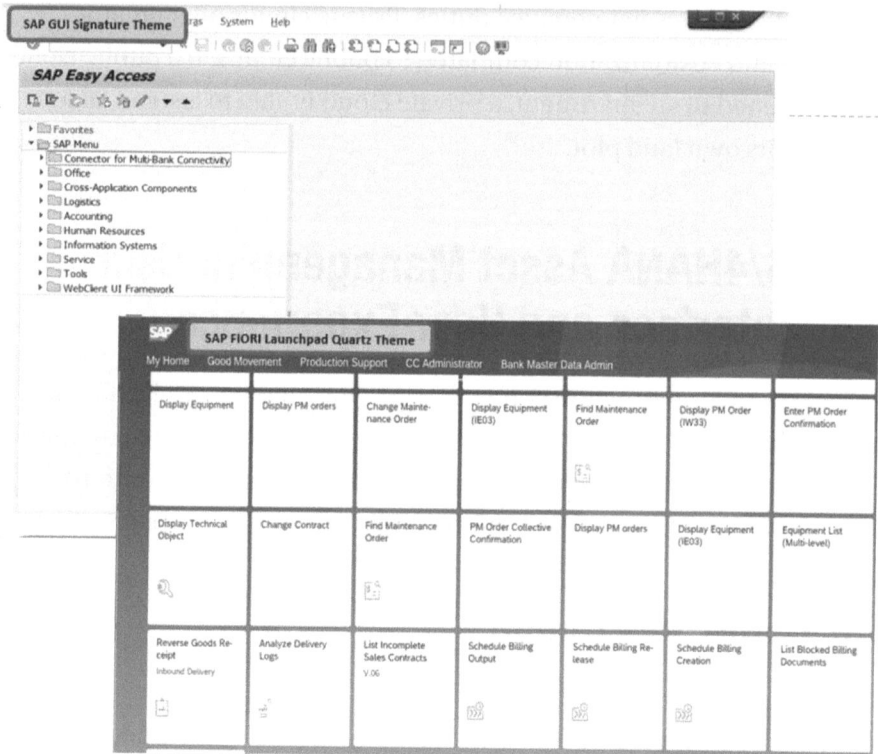

Figure 1-6. *S/4HANA user interfaces*

1.6.3. SAP Business Client

The SAP Business Client UI offers a unified landing point for various SAP business applications and technologies. Customers with S/4HANA on-premise deployment can use SAP Business Client to execute both SAP Fiori apps and SAP GUI for Windows transactions.

1.7. **Summary**

This chapter explored the features and functionalities of SAP S/4HANA, a cutting-edge business suite designed to revolutionize business processes. We discussed different ways to provision SAP S/4HANA, including on-premises and cloud deployments, ensuring organizations can choose the most suitable approach for their needs.

You gained insights into the comprehensive asset management capabilities within SAP S/4HANA, which includes maintenance planning, execution and completion, real-time tracking, and reporting and analytics. The details also shed light on the key functionalities offered by the S/4HANA Asset Management application. Finally, we examined the user-friendly interface and user experience the SAP S/4HANA Asset Management application provides, empowering users with an intuitive and efficient tool.

By harnessing the potential of SAP S/4HANA, businesses can optimize asset management processes, leading to improved productivity and enhanced decision-making capabilities.

CHAPTER 2

Asset Management Organizational Structure

In an efficient asset management application, a well-defined organizational structure is the foundation for successful strategies. This chapter delves into the important elements of organizational units and their significance in S/4HANA Asset Management. It navigates through a series of topics, each shedding light on a crucial aspect of this structure.

The following are some of the key topics covered in this chapter.

- Introduction to organizational units in S/4HANA
- Organizational units specific to asset management
- Maintenance plants
- Planning plants
- Centralized and decentralized planning plants
- Maintenance work centers
- Configuring an organizational structure in S/4HANA

© Rajesh Ojha and Chandan Mohan Jaiswal 2023
R. Ojha and C. M. Jaiswal, *SAP S/4HANA Asset Management*,
https://doi.org/10.1007/978-1-4842-9870-1_2

2.1. Introduction to Organizational Units in S/4HANA

Defining organization units is an elementary requirement to structure and organize various components of an organization's structure within S/4HANA. These organization units help the S/4HANA applications to represent the real-world functional structure of the organization, such as finance, logistics, sales, human resources, and others (see Figure 2-1). Each organization unit serves a specific purpose and is crucial in managing data and business operations within the S/4HANA application.

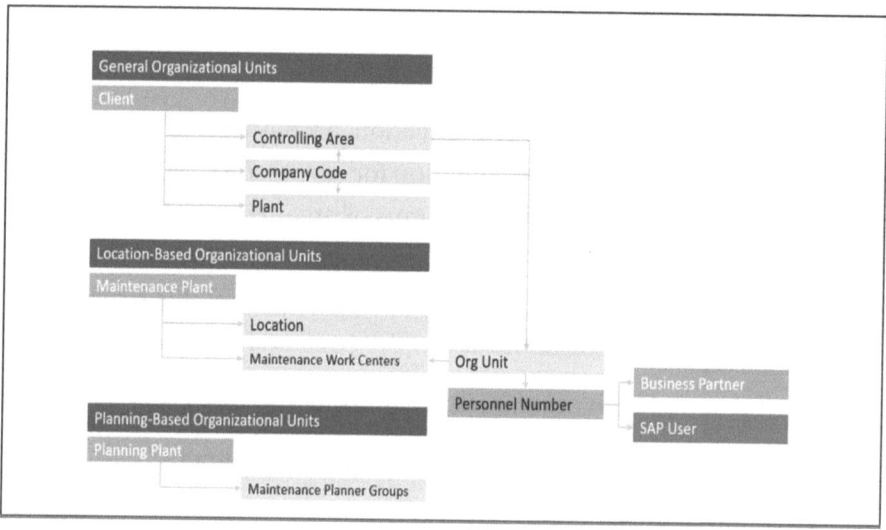

Figure 2-1. *Organizational units in S/4HANA*

The following are some of the most important organizational units in S/4HANA.

- **Client**: The client is the top-level entity among all organizational units; for example, it may represent a conglomerate of corporations comprising multiple companies. All the business applications within a client share a common database for data.

- **Controlling area**: A controlling area in S/4HANA is a way of grouping and organizing one or more company codes to enable centralized control and management accounting functions across multiple subsidiaries or business units within an organization.

- **Company codes**: A company represents an independent legal entity or a business unit within a corporate group. For instance, if you have a parent company with multiple subsidiaries, each subsidiary may be represented as a separate company code in S/4HANA.

- **Plant**: A plant is a physical location or facility where the production or stocking of goods occurs. It could be a manufacturing plant or a warehouse.

- **Maintenance plant**: Among the plants, those at which technical systems (machines, equipment, assembly lines) are installed and operationalized are called *maintenance plants*.

- **Planning plant (maintenance planning plant)**: Among the plants, those at which maintenance planning is performed are defined as *planning plants*. Maintenance planning (e.g., creating and managing preventive maintenance plans, work orders, schedules, and optimizing resources) is performed in a planning plant.

- **Maintenance work center**: The technical and human resources used for asset management are known as *work centers* (often called *workshops* in business) and are associated with a plant.

- **Organizational unit**: An organizational unit is one of the elementary objects in *human capital management* (HCM) used to construct an HCM organizational structure. An organizational unit can be linked with a work center in asset management, enabling the specification of different teams from both a logistical and an HR perspective.

- **Business partner**: A business partner is the functionality to create master data, such as persons, groups of persons, and organizations; examples include customers, vendors, and employees. Each employee (personnel number) has a mandatory business partner assigned.

2.2. Organizational Units Specific to Asset Management

In S/4HANA Asset Management, an organizational unit represents a fundamental element of the organizational structure within the system. It is a key building block for effectively organizing and managing assets, maintenance processes, and resources.

The following sections describe a few of the most important organizational units in S/4HANA Asset Management.

2.2.1. Maintenance Plants

A maintenance plant is an important organizational unit representing a physical location or facility where technical systems (machines, equipment, assembly lines) are installed and operationalized. For example, a plant where an automotive assembly line is set up, a machine commissioned and operationalized, or a vehicle used for transporting goods.

The following describes the characteristics of a maintenance plant, which comprises various locations, sections, and production facilities.

- **Locations**: Locations represent different parts/areas within a plant. For example, if a plant consists of a store, dispatch area, divisions, and parking area, each can be defined as a location within the plant. You can use location as an input parameter in various reports to filter your information, such as for the equipment master and functional location lists.

- **Plant section**: Maintenance plants are divided into production areas, with machines or sets of machines represented in the system as equipment or functional locations installed within these plant sections. Plant sections indicate the specific location for which a maintenance task was requested. The plant section can also designate a contact person for plant maintenance in production.

- **Work center**: The technical and human resources used for asset management are called work centers.

2.2.2. Planning Plants (Maintenance Planning Plants)

An organizational unit in asset management where planning activities are conducted is known as a planning plant (see Figure 2-2). These activities include creating and managing maintenance plans, work orders, schedules, and optimizing resources. These planning activities may originate from the planning plant itself, which can also serve as a maintenance plant, or they may come from other maintenance plants connected to this maintenance planning plant. The planner assigned to the planning plant is defined as the maintenance planner in customization. The planner group is an individual or group responsible for maintenance planning in a planning plant.

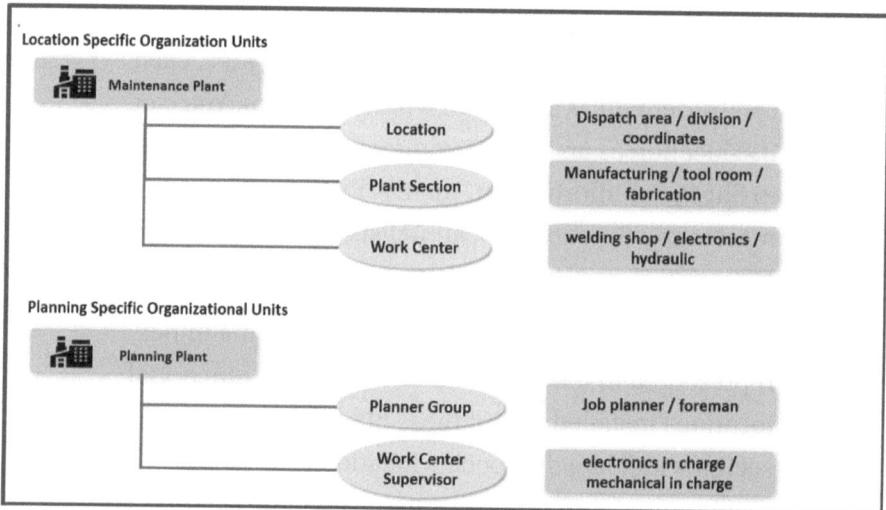

Figure 2-2. *Organizational units specific to location and planning*

2.3. Centralized and Decentralized Planning Plants

Based on the organization's asset maintenance planning structure—centralized or decentralized—S/4HANA Asset Management defines a planning plant as either centralized or decentralized.

2.3.1. Centralized Planning Plants

In a centralized planning plant, the planning activities for technical assets from multiple maintenance plants have been performed from one planning plant. In this case, the planning plant is responsible for all the maintenance plants, and planning activities are consolidated and managed in this planning plant. This central planning plant serves as a hub where maintenance plans, work orders, and schedules are created and coordinated for multiple maintenance plants within the organization. Components are stored and issued from the centralized planning plant. However, the maintenance work center can be set up in either the planning plant or the maintenance plant, depending on the location of the maintenance workshop. If the maintenance workshop is centrally available to serve all requesting maintenance plants, the work center needs to be set up with the planning plant. But in case a few requesting maintenance plants have their own locally available workshops, then for those workshops, the work centers need to be set up in their respective maintenance plants (see Figure 2-3).

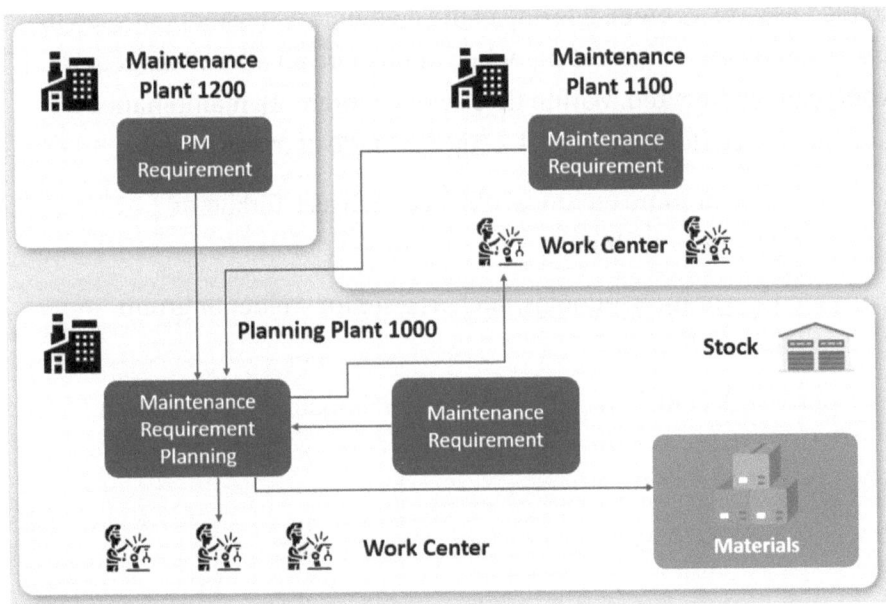

Figure 2-3. *Centralized planning plant*

2.3.2. Decentralized Planning Plants

In contrast, a decentralized planning plant allows each maintenance plant to manage its plan independently. In this scenario, each maintenance plant is also its own maintenance planning plant and carries out activities such as creating maintenance plans, work orders, and schedules for its technical assets. The work center processes the maintenance orders, and components are issued from the same plant.

2.4. Maintenance Work Centers

The technical and human resources used for asset management are defined as work centers. It is one of the elementary master data used in asset management that defines when and where an operation (repair work) can be executed. Within these work centers, all maintenance activities are performed. A work center is made of the following.

- One or more technical objects, such as a forklift or welding machine

- One or more persons, such as a senior welder or group of mechanics

These work centers (see Figure 2-4) are associated with a plant.

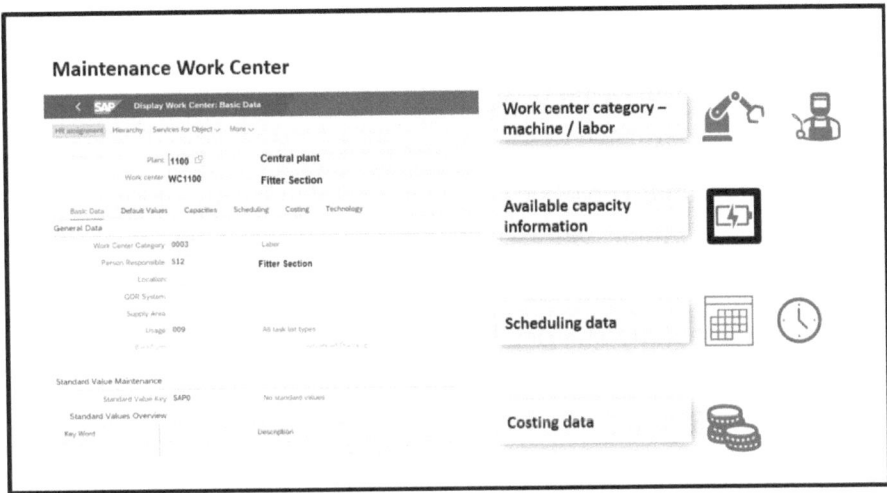

Figure 2-4. *Maintenance work center*

Maintenance work centers are maintained in various master data and used in several transactions.

- In the functional location and equipment master, they are responsible for maintenance work.

- In the maintenance plan's item, they carry out planned maintenance work.

- In the task list header section, they are responsible for maintenance work.

- In each operation of the task list, they execute the operation.

- They are responsible for maintenance work in the maintenance and repair order header section.

- In each maintenance and repair order operation, they execute the operation.

2.4.1. Work Center Sections and Information

The following explains a few of the important sections and data within these sections of a work center.

- **Basic data**: The Basic Data view/tab of the work center contains information such as category, description, the person responsible, usage, and standard value key.

- **Default data**: The Default Values view of the work center contains information for transferring a few of the default values, such as the control key, when the work center has been entered in an operation of a maintenance order. There is also an option to maintain a few more default values if required. By activating a specific business function, you can maintain default values for the following.

    - Contract and contract item

    - Vendor (supplier)

    - Purchasing organization

    - Purchasing group

    - Material group

- **Capacity data**: The Capacity view of the work center contains information related to the resource capacity of the work center, such as capacity category (machine/person), the number of resources available, working hours, and break hours. The available capacity may be limited based on time or arranged in shifts that deviate from the standard available capacity.

- **Scheduling data**: Based on the scheduling data, an operation's start and finish time (duration of an operation) are calculated based on lead times.

- **Costing data**: The Costing view of the work center contains important information related to the internal cost of the resources of the work center, such as the cost center with which the work center is associated, costing activity types (various rates based on resource type) and validity period (see Figure 2-5).

- **Personal data**: Along with capacity-related information, you can assign people (employees' personal data in the Human Resource application) or their positions in the work center.

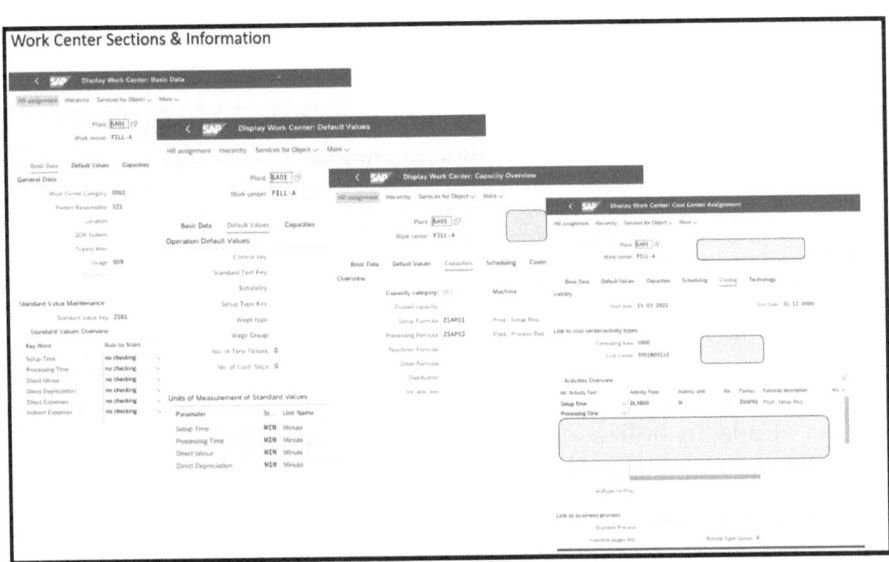

Figure 2-5. *A work center's various sections and information*

The relationship between work centers and other objects within the SAP S/4HANA system is established through maintenance work center links. The linkage with other objects, such as people, positions, and qualifications, connects and organizes various resources and information within the system. These links are maintained with a certain validity period. A work center can be linked with the following objects.

- People

- Positions

- Qualifications

- Cost centers

2.4.2. Work Center Capacity Planning

To utilize a work center's resources during the execution of maintenance work for an asset, the work center must have available capacity within a specific period as required. The system calculates available capacity in consideration of values from the following fields.

- Work center start and finish times

- Length of breaks (such as for lunch)

- Total number of unit capacity available (such as number of technicians)

- Capacity utilization percentage (number of hours per day available for working on planned activities)

The available capacity has two types (see Figure 2-6).

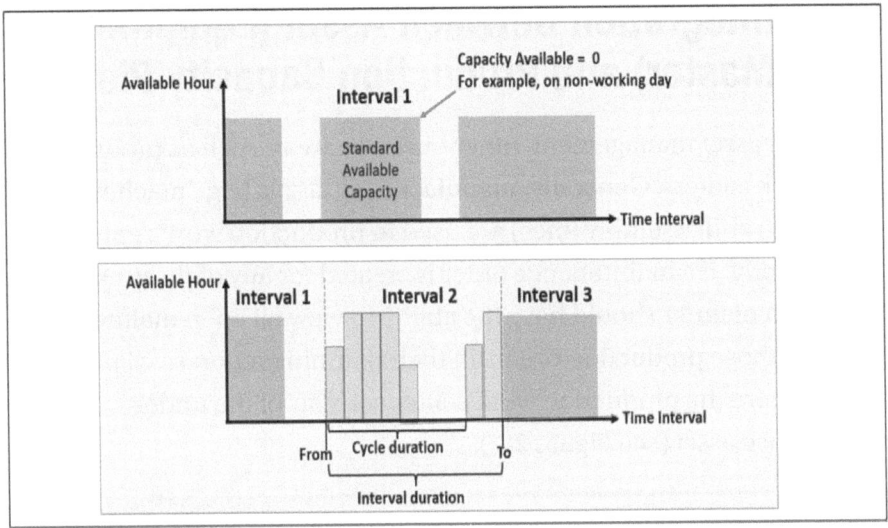

Figure 2-6. *Available capacity in the work center*

- **Standard available capacity**: The standard available capacity is calculated using values such as the work center's start time, end time, break times, and individual capacity number. The capacity is available without any time intervals (shifts). With this data, the system calculates the working time of the capacity. The capacity is available for the entire duration of the work center's operating time, which means an equal number of capacities are available throughout the period.

- **Intervals of available capacity**: Intervals of available capacity refer to the specified start, finish, and break times for work centers within a shift, along with validity periods for the available capacity. These timeframes indicate the capacity that can be utilized during specific validity periods. To define an interval of available capacity, you can use shift programs that apply to all work centers for a fixed time period.

27

2.4.3. Integration Between Asset (Equipment Master) and Production Capacity Planning

Along with asset management, other business areas, such as production, utilize work centers. Generally, manufacturing assets (e.g., machines, equipment, and assembly lines) are used in production work centers. Consequently, if a maintenance order is created for any of these assets, the production planner should have the ability to view all open maintenance orders for those production orders in their planning report (e.g., planning board), where the production work center consists of the under-maintenance asset (see Figure 2-7).

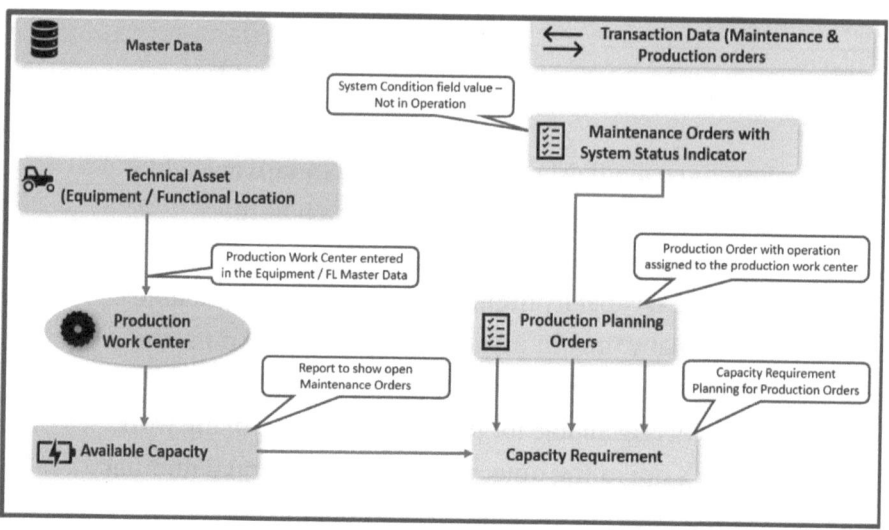

Figure 2-7. *Integration between asset (equipment master) and production capacity planning*

To accomplish this, the *production work center code* must be entered in the master data of the technical asset (equipment master's location data section).

The value for the System Condition field must be selected in the header section of the open maintenance and repair order.

The *system condition* assigned to a maintenance order determines whether the technical object entered into the maintenance order is under breakdown or in operation during the processing of the maintenance order. In the case of the system condition value "not in operation," it means that the technical object cannot be used for its intended purpose. In contrast, the value "in operation" means that the technical object is still available during maintenance order processing and can be used for its intended purpose.

2.5. Configuring an Organizational Structure in S/4HANA

Let's discuss the recommended steps for defining organization structure in S/4 HANA Asset Management.

After confirming the proposed organizational structure with the organization's business team, you need to add the required organizational objects, as described next, using the S/4HANA configuration.

Asset Management (also known as Plant Maintenance) is generally implemented only after other application modules, such as Finance (FI), Controlling (CO), and Procure-to-Pay (Materials Management (MM)), have been implemented. Hence, the majority of the organizational structure has already been configured. The configured structure needs to be reviewed from an asset management perspective, and additional organizational structures must be included for the asset management–specific areas.

After a detailed study, the organizational units required for asset management are set up in the system. Maintenance plants (which are normally already configured for the Procure-to-Pay module as logistic plants) and planning plants (also known as *maintenance planning plants*) are defined in the organizational structure (see Figure 2-8).

Organizational Levels – Planning and operational units of Asset Management must be implemented in the Enterprise Structure.			
Implementation Steps	Organizational object	Responsible Team	Integration with other applications
1. Analyze	Existing organizational structure	Asset Management consultant	All
2. Define	Maintenance plants, planning plants and maintenance work centers	Asset Management consultant	Logistic / Accounting
3. Assign	Planning plant to maintenance plant Create work centers in maint plant	Asset Management consultant	Logistic / Accounting

Figure 2-8. *Organizational levels*

In this step, the maintenance planner groups are created in planning plants, and maintenance work centers are created for maintenance plants.

Configuration is the process of customizing and setting up various functionalities and features of the SAP S/4HANA software according to the specific needs and requirements of an organization's business processes. Although the standard S/4HANA Business Suite comes preconfigured, which can be readily used for an organization's business process execution, a certain degree of additional settings and changes to existing settings are required. This allows system administrators and consultants to add and change settings and options within SAP S/4HANA to tailor the system as required. It is a prerequisite step in the SAP S/4HANA implementation process because it ensures that the software system is aligned with the organization's business processes. Among a comprehensive list of configuration options, you can configure a few of the most common and important objects (see Figure 2-9), described as follows.

- Organizational structures such as finance, logistics, asset management, and human resources

- Various document types, such as procurement, production, maintenance, and sales

- Mater data–related settings such as for equipment (assets), product (material), customer, and supplier

- Planning related objects such as material requirement planning and resource-related planning

- Various checks and validation-related settings for master data and business process transaction data

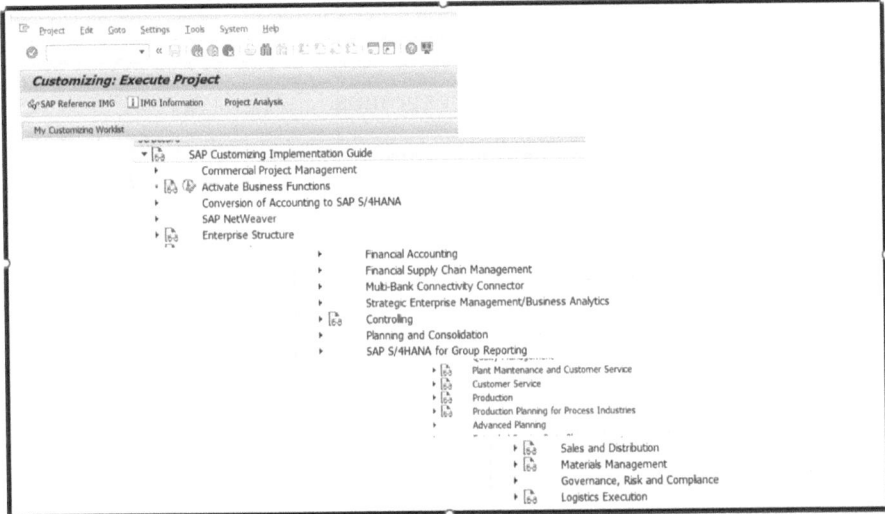

Figure 2-9. *A few of the important configuration nodes*

2.5.1. Configuring a Plant

Defining the plant is one of the elementary and important organizational unit configurations for logistic-related business processes such as procurement, inventory, and manufacturing. It is also a prerequisite for setting up the planning plant for asset management. Although all the required plants are configured for logistics before implementing asset management, typically configuring a plant involves the following steps.

1. Define the plant for logistics.

2. Identify the plant as a planning plant for asset management.

3. Define storage location in a plant.

4. Assign plant to company code.

5. Assign purchase organization to plant.

6. Assign planning plant to maintenance plant.

7. Define maintenance planner groups.

8. Update quantities/values for material types.

9. Define account determination.

10. Define plant parameters for inventory management.

11. Define plant parameters for consumption-based planning.

Let's discuss the navigation path to the Customization section. From the SAP Easy Access menu, navigate to Tools → Customizing. Double-click IMG → SPRO–Execute Project. Click the SAP Reference IMG button.

Table 2-1 lists important configuration paths related to the plant.

Table 2-1. *Configuration Path for Important Configuration Related to Plant*

Configuration Step	Configuration Path
Define Plant (not done by Asset Management Project Team)	Enterprise Structure → Definition → Logistics - General → Define, copy, delete, check plant
Define Planning Plant (configuration done by Asset Management Project Team)	Enterprise Structure → Definition → Plant Maintenance → Maintain maintenance planning plant
Assign Planning Plant (configuration done by Asset Management Project Team)	Enterprise Structure → Assignment → Plant Maintenance → Assign maintenance planning plant to maintenance plant
Define Planner Groups (configuration done by Asset Management Project Team)	Plant Maintenance and Customer Service → Master Data in Plant Maintenance and Customer Service → Technical Objects → General Data → Define Planner Groups
Storage Location (not done by Asset Management Project Team)	Enterprise Structure → Definition → Logistics - General → Define Location
Purchasing Organization (not done by Asset Management Project Team)	Enterprise Structure → Assignment → Materials Management → Assign purchasing organization to plant — / Assign standard purchasing organization to plant
Quantity and Value Update for Material Types (not done by Asset Management Project Team)	Logistics - General → Material Master → Basic Settings → Material Types → Define Attributes of Material Types
Account Determination (not done by Asset Management Project Team)	Materials Management → Valuation and Account Assignment → Account Determination → Account Determination Without Wizard → Group Together Valuation Areas

(*continued*)

Table 2-1. (*continued*)

Configuration Step	Configuration Path
Inventory Management (not done by Asset Management Project Team)	Materials Management → Inventory Management and Physical Inventory → Plant Parameters
Consumption-based Planning (not done by Asset Management Project Team)	Materials Management → Consumption-Based Planning → Plant Parameters → Carry Out Overall Maintenance of Plant Parameters
Split Valuation (Refurbishment) (not done by Asset Management Project Team)	Materials Management → Valuation and Account Assignment → Split Valuation → Activate Split Valuation — / Configure Split Valuation
Valuation of Goods (Refurbishment) (not done by Asset Management Project Team)	Production → Shop Floor Control → Integration → Define Valuation of Goods Received

For Asset Management the SAP S/4HANA offers the following configuration functionalities for organization objects.

1. **Mark a plant (logistic plant) as a planning plant.**
   In this configuration step, you must mark a defined logistic plant as the planning plant. If all the plants in the organizational structure are also maintenance planning plants, then you must mark each as a maintenance planning plant.

2. **Assign the planning plant to maintenance plants.**
   In this configuration step, you need to assign all the maintenance plants to their respective planning plants. This configuration step must assign

Every maintenance plant to a planning plant. A
maintenance plant can be assigned to only one
planning plant, but a planning plant can be assigned
to multiple maintenance plants.

A maintenance plant can be assigned to a planning plant in various
ways based on the actual organizational structure of asset management.

For example, a maintenance plant may be itself a planning plant
for the assets operationalized in the maintenance plant. Additionally,
a maintenance plant can be assigned to another planning plant.
Several maintenance plants can be assigned to one planning plant (see
Figure 2-10).

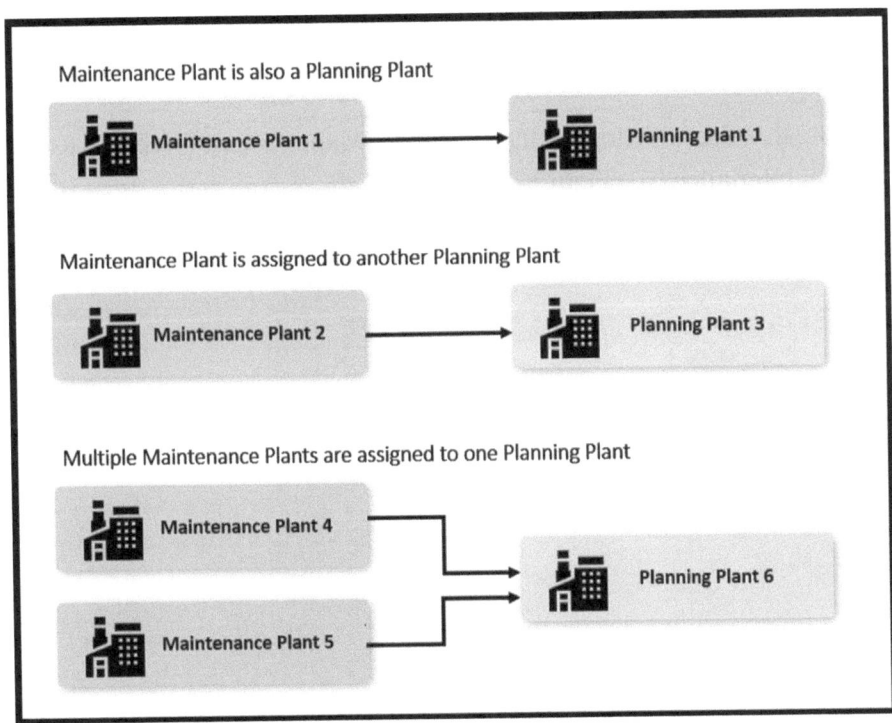

Figure 2-10. *Various relationships between maintenance plant and
planning plant*

Table 2-2 lists important configuration paths related to planning plant.

Table 2-2. *Configuration Path for Identifying Planning Plant and Assigning to a Maintenance Plant*

Configuration Step	Configuration Path
Identify Planning Plant (Maintenance Planning Plant)	SAP Customizing Implementation Guide → Enterprise Structure → Definition → Plant Maintenance → Maintain maintenance planning plant
Assign Planning Plant to Maintenance Plant	SAP Customizing Implementation Guide → Enterprise Structure → Assignment → Plant Maintenance → Assign maintenance planning plant to maintenance plant

Note that the node for configuring Asset Management is labeled as Plant Maintenance (see Figure 2-11).

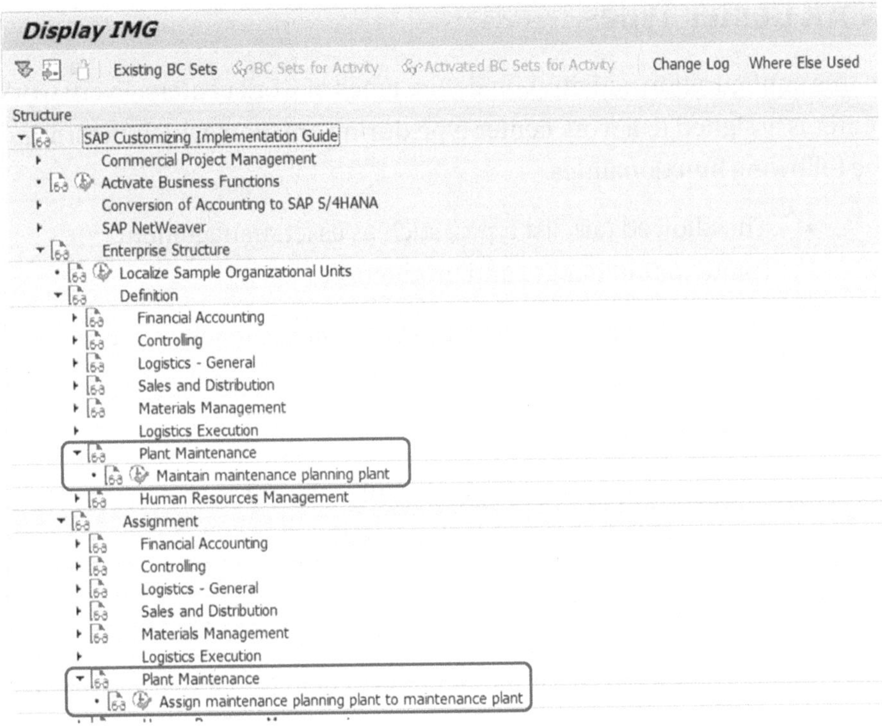

Display IMG

| Existing BC Sets | BC Sets for Activity | Activated BC Sets for Activity | Change Log | Where Else Used |

Structure

- SAP Customizing Implementation Guide
  - ▶ Commercial Project Management
  - Activate Business Functions
  - ▶ Conversion of Accounting to SAP S/4HANA
  - ▶ SAP NetWeaver
  - Enterprise Structure
    - Localize Sample Organizational Units
    - Definition
      - ▶ Financial Accounting
      - ▶ Controlling
      - ▶ Logistics - General
      - ▶ Sales and Distribution
      - ▶ Materials Management
      - ▶ Logistics Execution
      - Plant Maintenance
        - Maintain maintenance planning plant
      - ▶ Human Resources Management
    - Assignment
      - ▶ Financial Accounting
      - ▶ Controlling
      - ▶ Logistics - General
      - ▶ Sales and Distribution
      - ▶ Materials Management
      - ▶ Logistics Execution
      - Plant Maintenance
        - Assign maintenance planning plant to maintenance plant

Figure 2-11. *Configuration nodes for Asset Management organization objects*

2.5.2. Configuring a Work Center

Based on existing organizational maintenance workshops, you must create work centers in S/4HANA Asset Management for each workshop. Along with asset management, other business areas, such as production, use work centers.

A few functionalities and information items used in the work center can be configured according to the requirements.

Work Center Types

In this configuration activity, you define types of work centers. Every work center is assigned to a work center type during creation, which determines the following functionalities.

- The allowed task list types, such as asset management (plant maintenance) and production

- The screen sequence, such as basic data, capacity data, and costing view

- Field selection

- Whether changes to the work center should be documented with change documents

Field Selections

This configuration step determines whether a field is mandatory, optional, or display-only. You can configure fields in the following screen areas.

- Task lists

- Header screens

- Sequence screens

- Operation screens

Standard Value Keys

In this configuration step, you can define the standard value keys. The standard values represent planned values used for calculating the execution time of an operation in the maintenance task list.

Default Work Centers

In this configuration step, you can define a default work center for each work center type and plant.

Figure 2-12 shows the navigation path to the customization section.

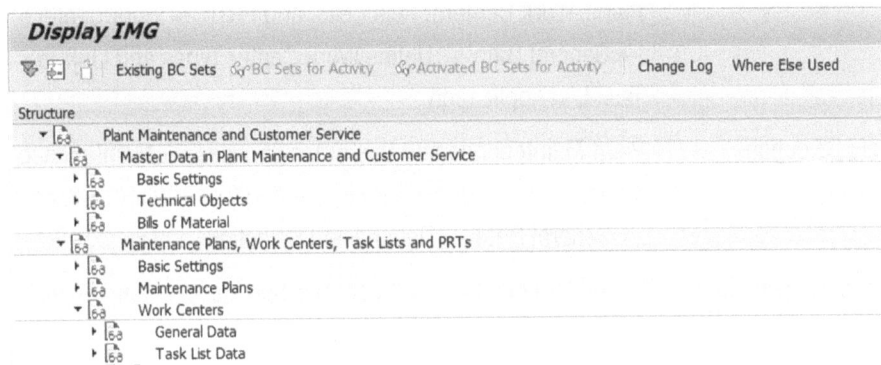

Figure 2-12. *Configuration path for work center functionalities and information*

From the SAP Easy Access menu, navigate to Tools → Customizing. Double-click IMG → SPRO–Execute Project. Click the SAP Reference IMG button.

Table 2-3 lists the configuration path for work center functionalities and information.

Table 2-3. *Configuration Path for Work Center Functionalities and Information*

Configuration Step	Configuration Path
Work Center Types	Plant Maintenance and Customer Service → Maintenance Plans, Work Centers, Task Lists, and PRTs → Work Centers → General Data → Define Work Center Types and Link to Task List Application
Field Selection	Plant Maintenance and Customer Service → Maintenance Plans, Work Centers, Task Lists and PRTs → Work Centers → General Data → Define Field Selection
Standard Value Keys	Plant Maintenance and Customer Service → Maintenance Plans, Work Centers, Task Lists and PRTs → Work Centers → General Data → Define Standard Value Keys
Default Work Center	Plant Maintenance and Customer Service → Maintenance Plans, Work Centers, Task Lists and PRTs → Work Centers → General Data → Create Default Work Center

Once all the prerequisite configurations for the work center have been checked and, if required, modified, you can create a new work center as master data. Typically, creating a work center involves the following steps.

1. Clarify which work center types you need.

2. Enter the work center descriptions.

3. Assign the keys of work center types for the screen sequence that serves as a reference.

4. Link the work center category to a task list application.

Next, let's discuss user interface for managing a work center.

SAP Fiori apps for work center maintenance are Create Work Center, Change Work Center, and Display Work Center.

For, SAP GUI-based transactions for work center maintenance, you can navigate to create, change, and display the work center (see Figure 2-13). From the SAP Easy Access menu, go to Logistics → Plant Maintenance → Management of Technical Objects → Environment → Work Centers → Work Center → IR01 – Create / IR02 – Change / IR03 – Display.

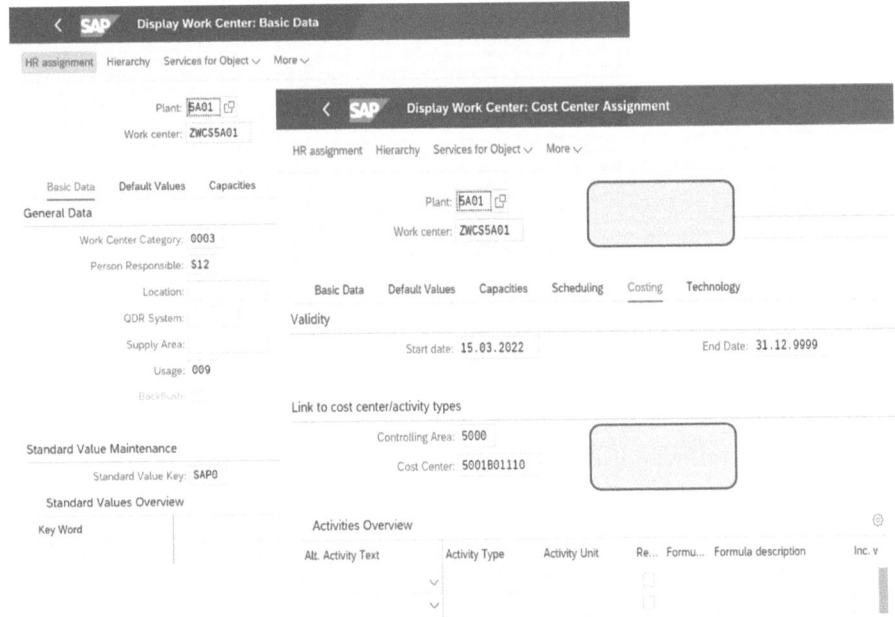

Figure 2-13. *Work center*

2.6. Summary

This chapter explained the role of organizational structures within an enterprise and the corresponding elements available in SAP S/4HANA. It delved into the organization structure specific to S/4HANA Asset Management, which includes maintenance plant, planning plant, and work center. Additionally, we explored various approaches to defining planning plants, such as centralized and decentralized planning plants.

The chapter also delved into the detailed process of configuring logistic plants, identifying a logistic plant as a planning plant, and assigning it to a maintenance plant. Finally, it detailed the configuration of a work center and how to create one.

This chapter provided valuable insights into the crucial elements that form the foundation of effective asset management within the S/4HANA system.

CHAPTER 3

Master Data for Asset Management

In the dynamic landscape of modern asset management, harnessing the power of SAP S/4HANA brings unprecedented opportunities, and at the core of this transformation lies in master data in SAP S/4HANA Asset Management. This chapter is a comprehensive guide to unlocking the potential of various master data components within the system.

Navigating through functional locations, equipment, classification, and linear asset management, this chapter lays the foundation for a structured approach to asset management, enabling organizations to efficiently allocate resources and streamline maintenance processes. It explores the intricacies of a bill of materials, serial numbers, measuring points, and counters, illuminating the critical role these elements play in enhancing operational accuracy and strategic decision-making.

As businesses seek to extract maximum value from their assets, mastering the intricacies of SAP S/4HANA becomes indispensable. This chapter acts as a beacon, guiding both novices and seasoned professionals through the rich landscape of master data management and facilitating the transition toward a more agile, informed, and responsive approach to asset management in the digital age.

© Rajesh Ojha and Chandan Mohan Jaiswal 2023
R. Ojha and C. M. Jaiswal, *SAP S/4HANA Asset Management*,
https://doi.org/10.1007/978-1-4842-9870-1_3

3.1. Functional Location

In SAP Enterprise Asset Management, a functional location is a physical or logical location within an organization where an asset or a group of assets are installed or used. It represents a specific place or area where maintenance activities are performed and is typically associated with a unique identifier in SAP.

Functional locations can be hierarchical in structure, allowing for the representation of different levels of detail. For example, you can have a functional location representing an entire building. Within that, you can define sub-functional locations for individual floors, rooms, or specific equipment within those rooms.

Functional locations serve as a foundation for asset management in SAP EAM. They enable you to track and manage maintenance activities at specific locations, such as inspections, repairs, and replacements. They also provide a structure for organizing and categorizing assets, allowing for better visibility and control over the maintenance processes.

By associating assets with functional locations, you can easily identify which assets are in a specific area and view their maintenance history, status, and other relevant information. This helps plan and execute maintenance tasks efficiently, optimize resource allocation, and ensure compliance with safety and regulatory requirements.

3.1.1. Structuring of Functional Locations

The following highlights the process of structuring technical objects.

1. Determine the assets that necessitate maintenance actions and those that require evaluation.

2. Select the appropriate organizational elements (functional location, equipment, assembly, material) for each asset and establish the structure.

3. Establish technical attributes (e.g., engine) and categories (e.g., locomotive classes). All technical attributes become accessible for an asset when associated with a class within a technical object (e.g., equipment), as shown in Figure 3-1.

4. Perform a configuration assessment on a prospective structure for a complex technical object arrangement.

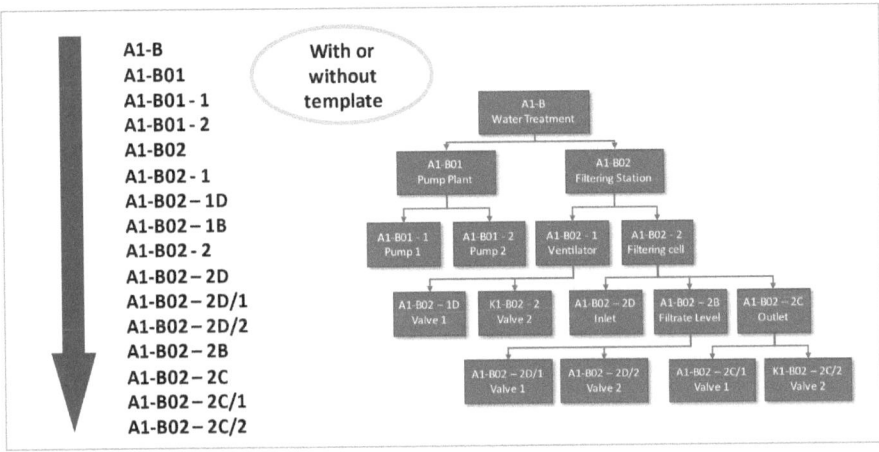

Figure 3-1. _List of entries for functional locations_

If the following conditions apply, utilize functional locations to organize your systems.

- You want to depict the structures of your company's technical systems based on functional criteria.

- Maintenance activities must be carried out on specific components of your technical system, and you need to maintain a record of this work.

- You need to gather technical data for specific parts of your technical system and analyze it over an extended period.

- You require monitoring of maintenance costs for specific components of your technical system.

- You intend to conduct analyses to determine the impact of usage conditions on the susceptibility to damage for installed equipment components.

Structure Indicator

Functional location labels (previously, numbers) are created using the structure indicator. The structure indicator consists of two input fields: Edit mask and Hierarchy levels (see Figure 3-2).

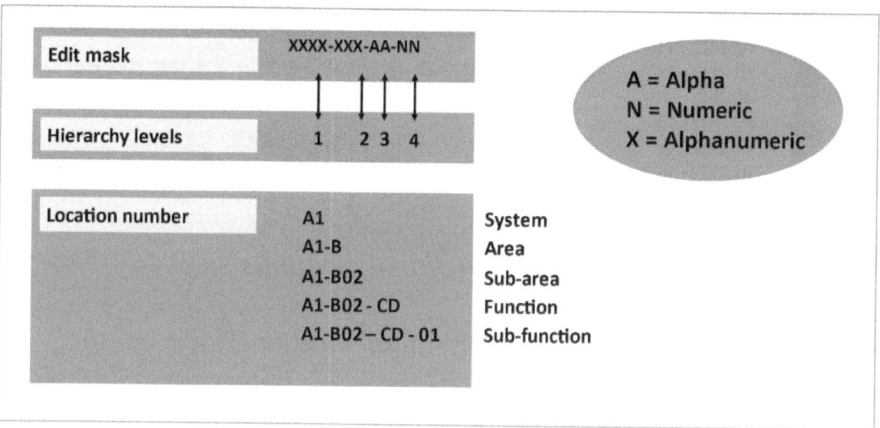

Figure 3-2. *Structure indicator*

An edit mask manages the allowable characters for labeling purposes (including letters, numbers, or both) and determine how these characters are grouped or separated. Hierarchy levels establish the endpoint character for each level and specify the maximum number of hierarchy levels that can be included in the structure. The functional location label is limited to 40 characters, as determined by the coding template's length.

You can configure functional location categories and structure indicators in IMG settings (see Figures 3-3 and 3-4).

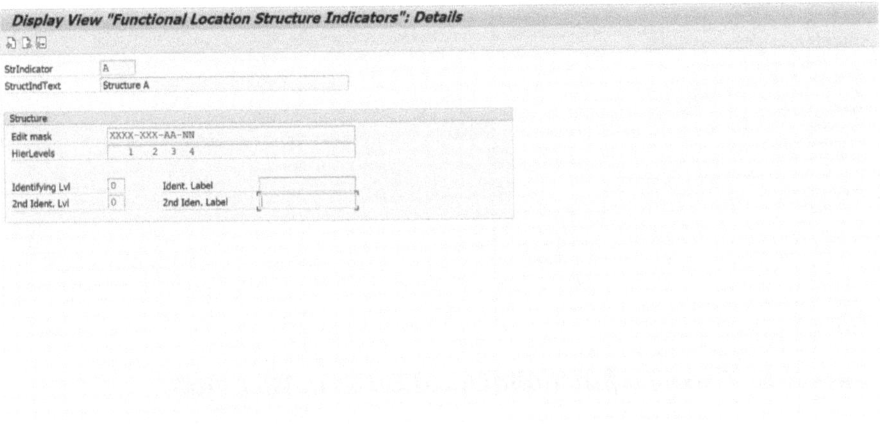

Figure 3-3. *Define structure indicator in IMG settings*

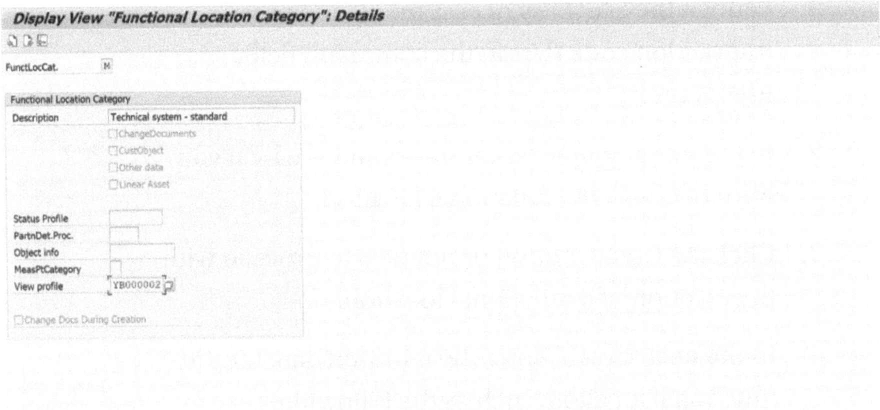

Figure 3-4. *Create a functional location category in IMG settings*

A functional location is created in transaction IL01 (see Figure 3-5). It is used for managing and tracking physical locations within an organization.

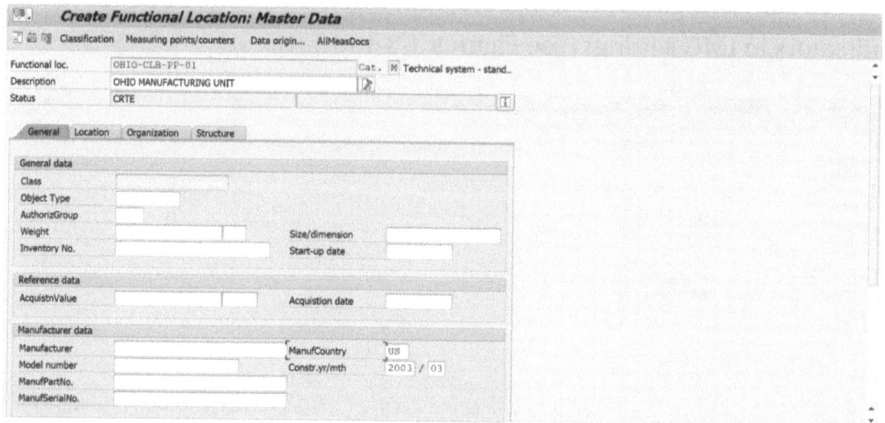

Figure 3-5. *Create a functional location as master data*

The following steps create a functional location in SAP using transaction IL01.

1. Launch the SAP Easy Access menu and enter transaction code IL01 in the command field. Press Enter.

2. In the initial screen, enter the Plant for which you want to create the functional location.

3. Click the Create button or press F5 to proceed with the creation of a functional location.

4. In the next screen, enter the required data for the functional location, such as the following.

    a. Functional location: Enter a unique identifier for the functional location.

    b. Description: Provide a meaningful description of the functional location.

    c. Planning plant: Enter the plant code if it differs from the one entered in the initial screen.

    d. Location: Specify the physical location details, such as building, room, or other relevant information.

5. You can also provide additional details for the functional location, such as maintenance plant, cost center, responsible work center, and other relevant data per your organization's requirements.

6. Once you have entered all the necessary information, click the Save button, or press Ctrl+S to save the functional location.

7. A message confirms the successful creation of the functional location, along with the newly assigned functional location number.

Reference Functional Locations

You can utilize a reference functional location to streamline the creation and administration of multiple similar functional locations within the system. Individual master records are used to define and manage these reference functional locations. It's important to note that reference functional locations do not represent actual physical locations; instead, they are assigned as references to real functional locations shown in Figure 3-6.

The reference functional location's master record encompasses information that applies to all functional locations linked to it. You only need to input location-specific data when generating functional locations using these reference locations, as the rest is inherited from the reference.

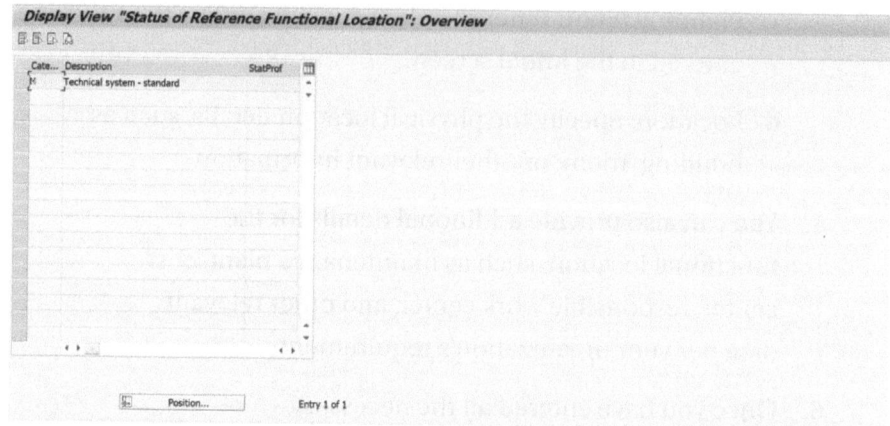

Figure 3-6. *Define reference functional location in IMG settings*

At the client level, the system manages master records for reference functional locations. Consequently, their labels are unique across the entire corporate group.

3.2. Data Transfer

When establishing a hierarchy with functional locations, various choices exist regarding data inheritance at lower levels.

The goal is to allocate specific core data at the highest feasible level and have it automatically propagate to numerous objects. This transfer can be centrally managed from the topmost functional location for all master record fields. Additionally (see Figure 3-7), you can disable data transfer at a specific hierarchy level and field (e.g., cost center), enabling individual maintenance of each field.

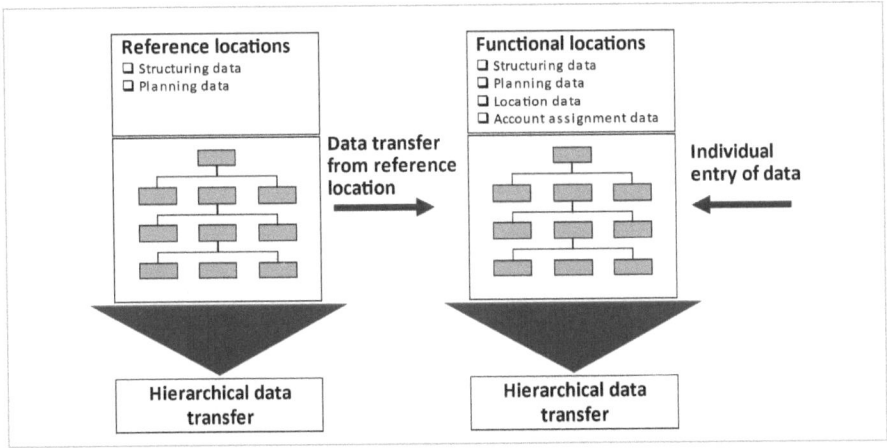

Figure 3-7. Data origin

3.2.1. Hierarchical Data Transfer

Functional locations and equipment hierarchies involve data transfer, following the following principles.

Starting from the Superior Object

For instance, data transfer occurs seamlessly when equipment is installed from a functional location. Fields in the equipment master record that already have entries remain unaffected and are regarded as individually maintained. Any fields without entries automatically inherit data from the functional location.

Starting from the Subordinate Object

If installation is initiated from the equipment master record, data transfer can be controlled using the Installation with Data Transfer function. This allows you to specify which fields should be maintained individually, independent of the functional location, and which fields should be copied from the functional location (see Figure 3-8).

51

To modify the data inheritance options for each field within the functional location, the Data Origin function can be utilized. For equipment, this switch can be made by dismantling the equipment and subsequently performing installation and configuration using the Installation with Data Transfer function.

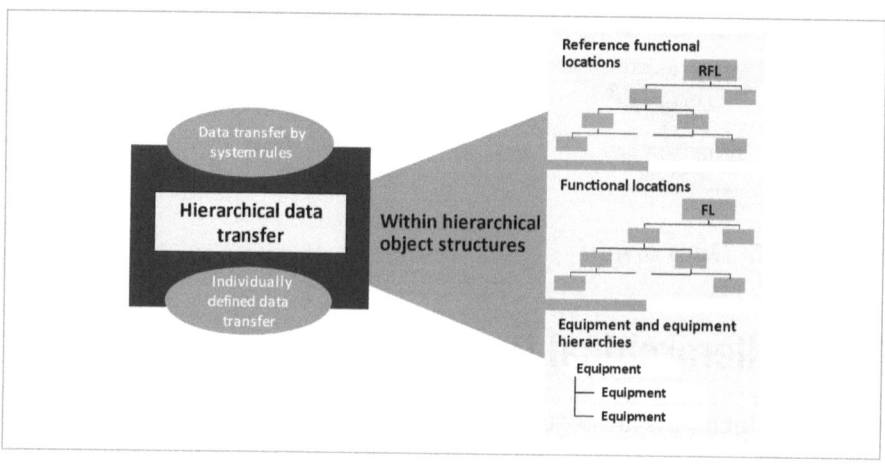

Figure 3-8. *Hierarchical data transfer*

3.2.2. Horizontal Data Transfer

Using reference functional locations, you can define category-specific data for each asset category and transfer this data to the corresponding functional locations, equipment, and associated subequipment (see Figure 3-9).

For instance, let's consider the responsible work center for the Electric Motor functional location that changes across multiple clarification plants. In such a scenario, the employee responsible for managing the master data would modify the master record for the Electric Motor reference functional location and save the changes. The system automatically updates all functional locations linked to this reference location and the equipment installed at those locations. A message is then generated, notifying the employee of the number of functional locations and equipment to which the data has been transferred.

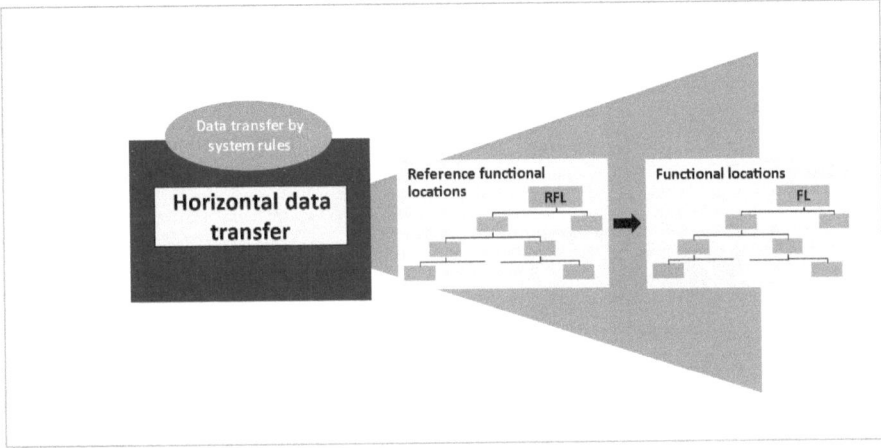

Figure 3-9. *Horizontal data transfer*

3.2.3. Data Origin

Regarding reference functional locations and functional locations (see Figure 3-10), you can view the data source in the master data fields. There are two display options available, as outlined next.

- Individual display: This option allows you to view the data origin for each field separately.

- Overview display: With this option, you can obtain a comprehensive overview of the data origin.

For reference functional locations and functional locations, you can specify the data source for each master data field. In the case of reference functional locations, you can determine whether the data should originate from a higher-level location in the structure or be maintained individually in the master record. Similarly, for functional locations, you can specify

53

whether the data should come from a superior location in the structure, a reference functional location, or be maintained individually in the master record.

For equipment, an overview display (in the form of a data origin list) is available for each tab's page. However, there is no field-specific display or option to make changes at the field level.

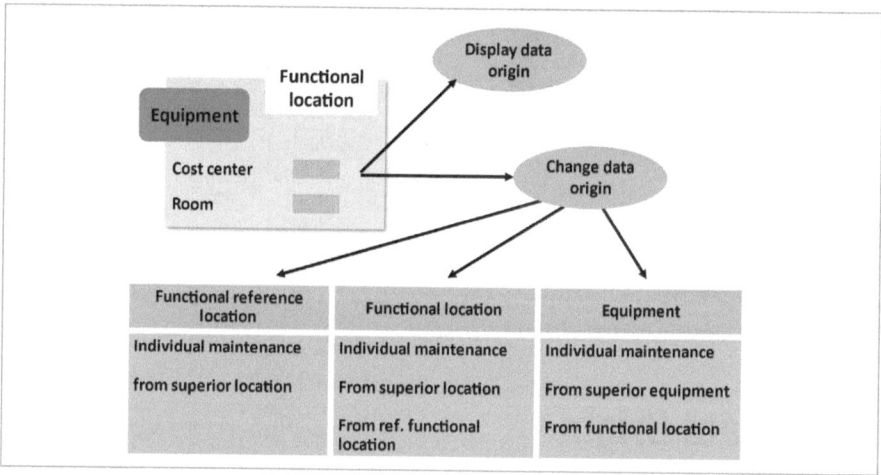

Figure 3-10. *Data origin*

3.3. Defining Alternate Labeling

In the case of alternative labeling for functional locations, you can modify the labels assigned to functional locations.

This modification can involve either "renumbering" the functional location by changing its primary key or creating an additional view that presents the object with a different numbering system. When the primary label is altered, you are prompted to determine whether the renumbered functional location should be classified under a new hierarchy, provided it is available.

Note   To enable alternative labeling, you need to activate it through customizing because it is not active by default in the standard system. It is advisable to thoroughly test this functionality before actual usage, as it can impact system performance.

For performance optimization, executing the report RI_IFLOT2IFLOD (via transaction code SE38) is recommended after activating alternative labeling.

While it is possible to deactivate alternative labeling, it's important to note that the system is not reset to its initial state before activation. In certain situations, deactivation may result in decreased system performance.

Alternative labeling of functional locations (see Figure 3-11) offers distinct labeling systems within the same hierarchy. This feature enables scenarios where different entities employ separate labeling systems, such as the system manufacturer and the customer.

It's important to note that alternative labeling systems, identified by an internal structure indicator, do not impact the object's structure. The primary labeling system determines the structure.

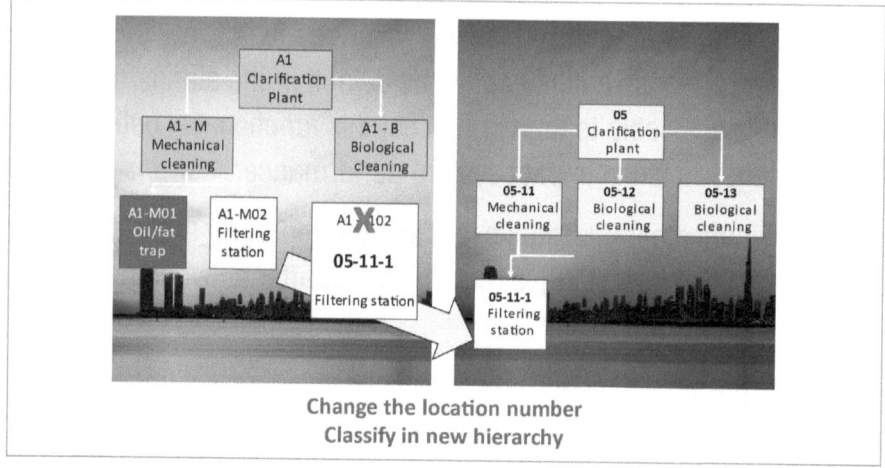

Figure 3-11. *Alternative labeling of functional locations*

The definition of the labeling system takes place in the Customizing settings. The selection of the appropriate labeling system is managed through a user profile, which enables the creation and activation of necessary views to accommodate the desired labeling system.

3.4. Standard Fiori Apps for Functional Locations

The following are some standard Fiori apps related to functional locations in SAP.

- Display Functional Location (App ID: F1692): This app allows you to view detailed information and characteristics of a functional location.

- Manage Functional Locations (App ID: F1076): With this app, you can create, edit, and delete functional locations. It provides comprehensive maintenance capabilities for functional locations.

- Search Functional Locations (App ID: F1993): This app enables you to search for functional locations based on various criteria, such as location ID, description, or hierarchy level.

- Change Functional Location Hierarchy (App ID: F1701): This app allows you to modify the hierarchy structure of functional locations by moving them within the hierarchy or creating new levels.

- Create Measurement Documents for Functional Locations (App ID: F1650): This app lets you record and manage measurement documents associated with specific functional locations.

- Manage Maintenance Orders for Functional Locations (App ID: F1705): With this app, you can create, edit, and manage maintenance orders related to functional locations.

These are just a few examples of standard Fiori apps available for functional locations in SAP. The availability and specific functionality of these apps may vary based on the SAP version and configuration.

Figure 3-12 shows Create Technical Object App.

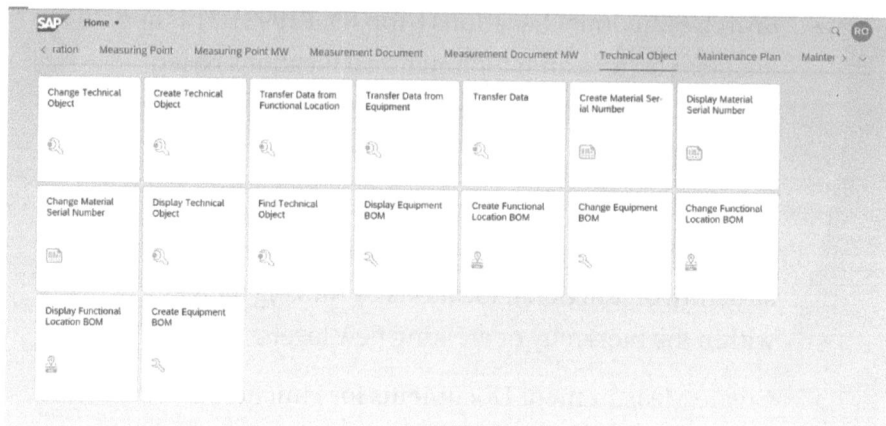

Figure 3-12. *Technical Object in Fiori apps*

3.5. Customizing Settings for Functional Locations

This section describes the Customizing settings for functional locations (see Figure 3-13 and Table 3-1).

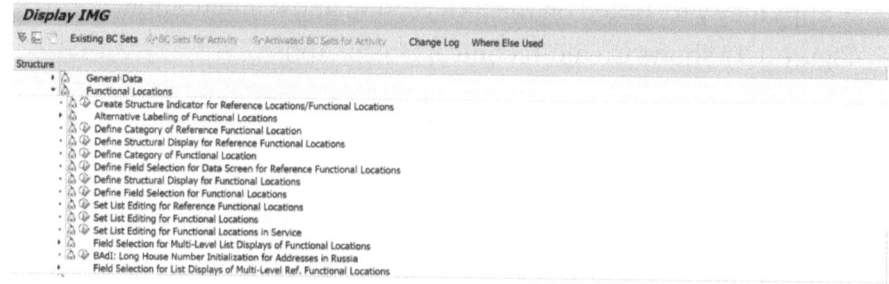

Figure 3-13. *IMG settings for a functional location*

Table 3-1. *IMG Settings for Functional Location*

Field Name or Data Type	Menu Path
Alternate Labeling	Plant Maintenance and Customer Service → Master Data in Plant Maintenance and Customer Service → Technical Objects → Functional Locations → Alternative Labeling for Functional Locations
List Editing	Plant Maintenance and Customer Service → Master Data in Plant Maintenance and Customer Service → Technical Objects → Functional Locations → Set List Editing for Reference Functional Locations Plant Maintenance and Customer Service → Master Data in Plant Maintenance and Customer Service → Technical Objects → Functional Locations → Set List Editing for Functional Locations
View Profile	Plant Maintenance and Customer Service → Master Data in Plant Maintenance and Customer Service → Technical Objects → General Data → Set View Profiles for Technical Objects
Structure Indicator	Plant Maintenance and Customer Service → Master Data in Plant Maintenance and Customer Service → Technical Objects → Functional Locations → Create Structure Indicator for Reference Locations/Functional Locations
Functional Location Category	Plant Maintenance and Customer Service → Master Data in Plant Maintenance and Customer Service → Technical Objects → Functional Locations Define Category of Functional Locations

3.6. Summary

In this unit, you learned that SAP functional location is a feature within the SAP Enterprise Resource Planning (ERP) system that allows organizations to manage and track physical assets or equipment. It is a unique identifier for a specific location where a functional object or equipment is installed or maintained.

Functional locations in SAP provide detailed information about an asset's physical characteristics, technical data, and maintenance history. They enable businesses to efficiently manage maintenance activities, schedule repairs, and track equipment performance.

Using SAP functional location, organizations can assign responsible personnel, track costs associated with maintenance, and plan preventive maintenance tasks. It offers a hierarchical structure to organize functional locations, allowing businesses to group assets based on various criteria such as location, department, or equipment type.

By integrating functional location data with other SAP modules, such as Plant Maintenance (PM) or Enterprise Asset Management (EAM), businesses can streamline their asset management processes, improve operational efficiency, and make informed decisions regarding maintenance and resource allocation. Overall, SAP's functional location serves as a central repository for managing and monitoring physical assets within an organization.

3.7. Equipment Master

In SAP EAM, an equipment master record is created to represent a specific piece of equipment within an organization. The equipment master record is a central repository of information related to that equipment, including its characteristics, maintenance history, location, and other relevant data.

The following are the steps to define an equipment master in SAP EAM.

1. First, you need to define an *equipment type* in the system. An equipment type represents a category or group of equipment with similar attributes or characteristics.

2. Once the equipment type is defined, you can create an *equipment class*. An equipment class further categorizes equipment types based on shared attributes and properties.

3. With the equipment type and class set up, you can proceed to create an *equipment master record*. This involves providing details such as equipment number, description, manufacturer, model, serial number, location, installation date, and other relevant information.

4. Depending on your organization's requirements, you can specify additional attributes and characteristics for the equipment. These attributes could include technical specifications, maintenance plans, warranty information, or other custom fields.

5. If the equipment has *dependencies* or *relationships* with other objects, such as functional locations or other equipment, you can establish these connections in the equipment master record.

6. You can define *maintenance plans* and *schedules* for the equipment, including planned maintenance tasks, frequency, and associated resources.

7. If specific measurements or readings are associated with the equipment, you can define *measurement points* for tracking and recording these values.

8. You can *attach documents*, manuals, diagrams, or other relevant files to the equipment master record for easy reference and access.

9. Define appropriate *authorization roles* and *access rights* to ensure that only authorized personnel can view and modify the equipment master record.

By following these steps, you can define and maintain equipment master records in SAP EAM, which provide a comprehensive overview of the equipment's details and enable effective management and maintenance throughout its lifecycle.

3.8. Define Equipment and Its Categories

The project team aims to utilize equipment to depict significant and valuable individual machinery components, such as pumps and motors. Equipment is employed to establish a historical record for objects related to plant maintenance or customer service.

Equipment can represent stand-alone objects, such as tools or vehicles, or denote assembled units at functional locations, such as pumps, motors, or gears (see Figure 3-14).

Figure 3-14. *Equipment master in SAP*

The equipment category serves as the primary assignment criterion. It is utilized to classify, describe, and manage fundamentally different objects, such as machines, fleet objects, production resources/tools, and customer equipment.

3.9. Equipment and Functional Location Relation

The combination of a hierarchical system or object structure, along with the autonomous object view, facilitates the representation of equipment installation at a functional location.

In the system, the installation location of an object or aggregate and the installed object can be represented separately. This differentiation allows distinct tracking of damages, enabling documentation of whether the damage is related to the conditions at the installation location (functional location) or caused by issues with the installed object (equipment).

The cause of the damage can be attributed and updated to either the functional location or the specific piece of equipment.

Usage times can be recorded from the equipment view (where the equipment was installed) and the functional location view (which equipment pieces were installed at this functional location).

3.9.1. Equipment Hierarchy

If functional locations are not supported in your company or individual movable aggregates possess a complex structure, equipment hierarchies can be utilized.

The equipment hierarchy operates differently from a functional location, as it lacks a structure indicator that establishes relationships between equipment numbers within the hierarchy shown in Figure 3-15. Additionally, in a Plant Maintenance Information System (PMIS), costs cannot be aggregated at the top level of equipment for the underlying hierarchy.

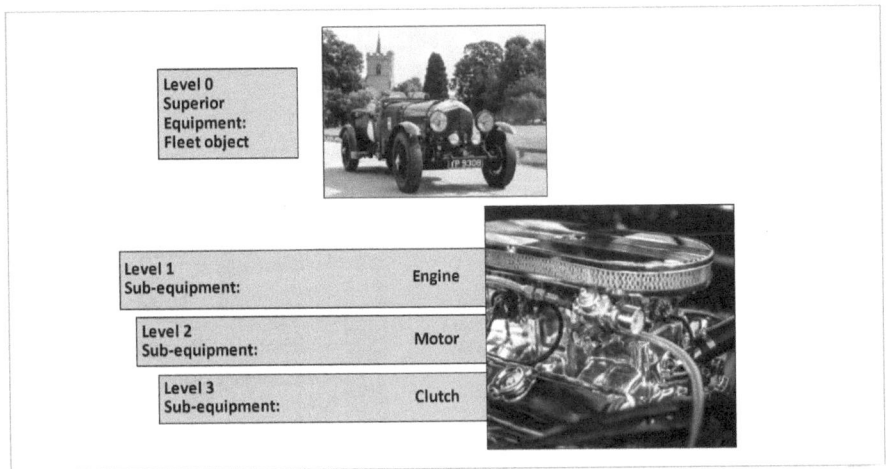

Figure 3-15. *Equipment hierarchy*

Equipment hierarchies can be structured according to specific requirements. Unlike installations at functional locations, there are no specific installation specifications for adding another piece of equipment. This means that equipment pieces not installed at a functional location can be installed within another piece of equipment.

Data transfer from superior equipment to subordinate equipment functions similarly to data transfer from a functional location to equipment.

3.9.2. Equipment Hierarchy with Functional Location

Equipment hierarchies can be installed at functional locations, just like individual pieces of equipment. The transfer of data follows the same principle as between functional locations and equipment or between functional locations and equipment hierarchies (see Figure 3-16).

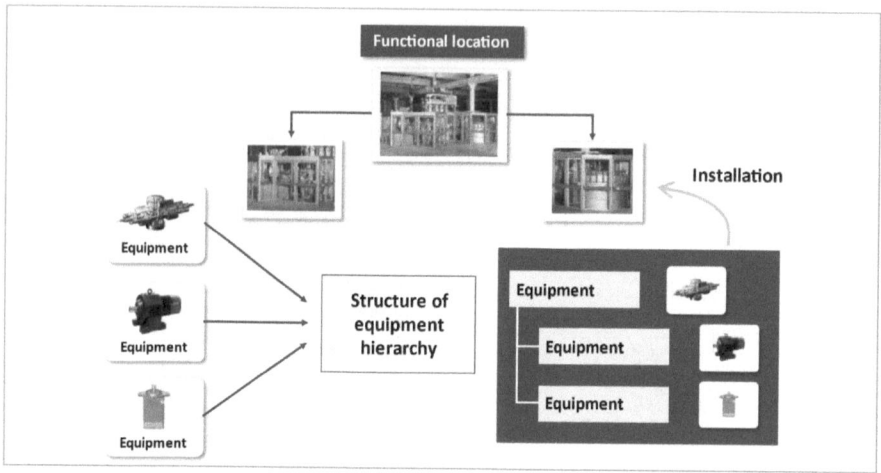

Figure 3-16. *Equipment hierarchy and functional location*

When equipment usage periods are displayed at the functional location level, the pieces of equipment within the hierarchy and the entire hierarchy can be viewed at the top level.

In the Customizing settings, it is possible to create equipment categories that generate change documents. These change documents serve to document any modifications made to the equipment master records. Each change is recorded in a separate document and assigned a unique document number.

3.9.3. Equipment Categories

To organize equipment categories, you can utilize the object type, which applies to functional locations as well. For instance, if the equipment type is a fleet object, the object types can include cars, trucks, motorbikes, and more.

Moreover, apart from structuring based on equipment and object type, it is possible to incorporate special views into the master record. This can apply to production resources/tools and fleet objects.

The following describes production resources and tools.

- A production resource/tool (PRT) is a portable operational resource utilized in Plant Maintenance and Productions, such as tools or measuring devices.

- PRTs can be classified as equipment, materials, or documents, and they are assigned to specific operations within the maintenance order.

- In Maintenance Processing, you can perform status checks, availability checks, and review usage overviews for PRTs.

- PRTs possess an additional view, namely the PRT Data view, where you can input details about task list usage or set the status of the PRT to Locked.

A vehicle within SAP ERP is represented by an equipment master record containing vehicle-specific data.

To create an equipment master record with vehicle-specific data, you can utilize transaction code IE31.

The display of tab strips and fields can be configured based on the combination of equipment category and vehicle type.

For modifying and viewing vehicles in list editing, transaction codes IE36 and IE37 are accessible. These transactions enable the use of vehicle-specific selection criteria.

Within these transactions, you can perform the following actions.

- Enter measurement documents for fuel consumption and distance

- Record the consumption of operating supplies by creating a material document

In SAP ERP 6.0 (LOG_EAM_PAM business function) enhancement package 3, the Pool Asset Management feature becomes available. This functionality facilitates requirements planning for vehicles within an internal car fleet.

3.10. Status Management and Partner Assignment

Status management can be employed to define and ascertain the availability of a technical object for specific operations. Two types of statuses exist.

- System status

- User status

System statuses are internally assigned by the system for specific business transactions within the general status management. They indicate whether a particular business transaction has been executed for a technical object and determine which subsequent business transactions can be performed based on the assigned status.

As system statuses are automatically set by the system during specific business transactions, users are unable to directly alter them and can only view these statuses.

User statuses are determined by the system user and configured within a status profile. This status profile allows you to control the permitted business transactions corresponding to the relevant system statuses.

Regarding business operations, if you possess the necessary authorization, you can assign and remove user statuses that have been defined in the system's customization settings. By assembling the desired statuses into a status profile, you can then assign this profile to the object category, thereby extending it to the equipment category.

3.10.1. Partners

Partners (business partners, contact persons) are internal or external organization units that can be involved in the development of maintenance or service processes.

Using customer exit ICSV0008, you can utilize non-standard partner types. The standard system supports the following partner types.

- Customer
- Vendor (external)
- Contact person
- Personnel number
- SAP users (internal)
- Job
- Organizational unit

These partner types can be further categorized based on partner functions, which determine the rights, obligations, responsibilities, and tasks of partners during the processing of a business transaction. In most cases, a partner fulfills predefined functions, requiring only one master record for that partner. However, there are instances where partner functions are divided among different companies and their subsidiaries.

The partner determination procedure is a grouping of partner functions that defines whether partners are allowed for a specific object and which partner functions can be performed in business transactions. This procedure can be assigned to the object. When a notification or order is created for a technical object with a partner, the system automatically transfers the partners from the technical object to the notification or order.

3.10.2. Partner Transfer

When posting goods issues during a delivery, the partner data is automatically transferred from the delivery note to the master record for the respective serial number. The system takes the partner data from the sales and distribution (SD) document and copies it to the serial number's master record. If there is already partner data specified in the master record, it gets overwritten by the data from the SD document. However, this overwriting of partner data only occurs for the data that has been flagged with an indicator in the Customizing settings.

3.10.3. Address Management

The following objects are linked to address management.

- Functional location

- Equipment

- Notification

- Order

- Components for non-stock material (object)

You can assign an address to all technical objects. The addresses of technical objects, notifications, and orders are synchronized shown in Figure 3-17.

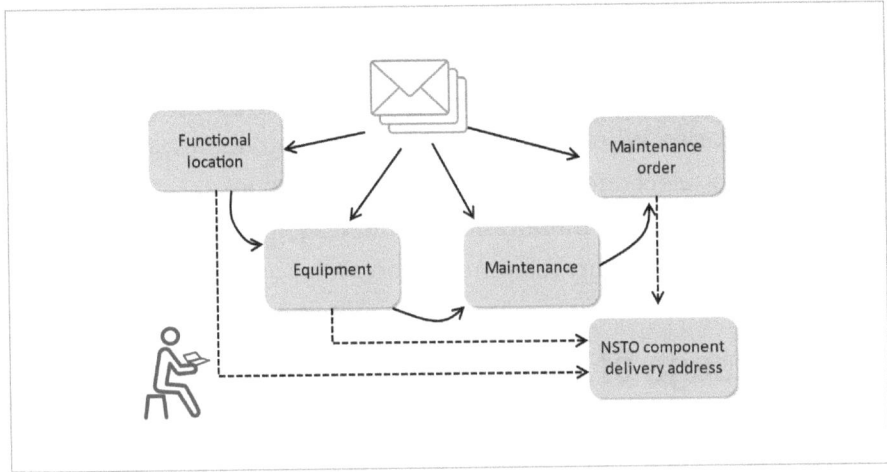

Figure 3-17. *Central address management*

3.11. Defining Warranties

A warranty is a contractual commitment to the customer, where services are provided either partially or entirely at no cost for a specified duration or for the lifespan of a specific device.

Before creating a warranty record, it is necessary to establish the intended usage of the warranty. This requires identifying the warranty type. The warranty type is associated with an internal warranty category that determines whether the company, on behalf of which the system user operates, acts as the warranty or the guarantor.

Within SAP, the following warranty types are distinguished from the perspective of the system user.

- Manufacturer warranty (inbound warranty)

- Vendor warranty (inbound warranty)

- Customer warranty (outbound guarantor)

SAP already provides a set of predefined entries in standard tables. In Customizing, it is possible to define additional warranty types as needed for your company and assign each warranty type to one of the two warranty categories. When creating a warranty record for a system user, it is mandatory to select one of the existing warranty types.

Warranty records can initially be generated as master warranties independently, without any reference object.

3.11.1. Usage of Warranty

The structure of a master warranty is as follows.

- The warranty header includes the warranty type, descriptive text, and classification option.

- The warranty items encompass the services covered or not covered by the warranty. These services can be described using text, service numbers, or material numbers. Configurable services can have assigned values within the warranty.

- Each warranty item can have defined warranty counters, which serve as validity criteria based on a time or a performance counter. The warranty counters can be linked with an AND logic (both conditions must be met) or an OR logic (at least one condition must be met).

A master warranty can be assigned to multiple pieces of equipment and can be linked to functional locations, equipment, and serial numbers.

The system supports the assignment of two warranty types per object. For each object, both a customer warranty and a vendor warranty can be assigned. In case of any damage, both warranties are verified.

- **Warranty**: The warranty is provided by an external party, such as a vendor or manufacturer. The party receiving the warranty is the warranty.

- **Guarantor**: The warranty is internal and is extended to the customer. The party providing the warranty is the guarantor.

3.12. Equipment Master and Asset Master Integration

To link pieces of equipment to an asset, you have the option to specify an asset number in the equipment master record, thereby establishing integration with asset accounting.

If the integration between asset accounting and EAM is enabled in Customizing, equipment can be automatically created when a technical system is created. Asset class settings play a role in distinguishing various mapping aspects, such as the types of equipment utilized.

Alternatively, you can configure the automatic creation of an asset when a piece of equipment is generated, or you can specify that changes made to the equipment master record should also reflect in the asset master record.

Integration can be implemented either through direct synchronization or workflow processes.

To ensure data consistency between assets and equipment, it is possible to define that fields in the equipment master record must also be updated when modifications are made to the asset master record.

3.13. Equipment Master Classification

In SAP, classification is a cross-application function utilized not only by Plant Maintenance but also by other applications.

One of the key purposes of classification is to assign detailed features to a technical object that cannot be accommodated within the fields of the SAP standard system's master record.

Additionally, classification offers an alternative search option where objects can be located based on their characteristics and associated features. For example, a search could be performed for objects with a power requirement of 2000 watts and a lifting height of 10 meters, among other criteria.

Within order processing, classification enables the search for spare parts that match specific features, such as finding a pump with identical attributes. The system leverages the classification data to conduct the search, and in the results list, a component can be selected and copied into the order.

The primary function of a classification system is to describe objects using characteristics and group similar objects into classes, facilitating their classification and improving their searchability.

The structure of a classification system involves three key steps.

1. Feature definition: Features of an object are described by utilizing characteristics, which are centrally created within SAP.

2. Class creation: SAP generates classes where the characteristics are assigned either during the class creation process or in a subsequent step. In the context of classification in SAP EAM, a class represents a category or group that helps organize and structure data related to assets, equipment, materials, or other objects within an organization.

3. Object assignment (classification): Once the necessary classes for classification are created, objects can be assigned to these classes. The objects are described by employing the characteristics associated with the respective class.

3.14. Characteristics Inheritance

Characteristic inheritance is the mechanism through which a characteristic and its values are propagated to all subordinate classes within a class hierarchy. The subordinate classes themselves do not contain the characteristic directly. This feature allows the definition of a central characteristic that is required in all subordinate classes. By assigning it only once to a superior class, there is no need for explicit assignment to each individual class.

When establishing a class hierarchy, characteristics from all superior classes are inherited by the lower classes. Consequently, characteristics that have not been explicitly assigned to a class but originate from superior classes are also displayed on the valuation screen.

3.14.1. Technical Object Classification

The master record of a technical object (functional location, equipment) allows its assignment to one or multiple classes.

When assigning the technical object to multiple classes, it is possible to define a standard class.

Once the technical object is assigned to a class, all the characteristics associated with that class become accessible within the object. Each object can have its own values for these characteristics shown in Figure 3-18.

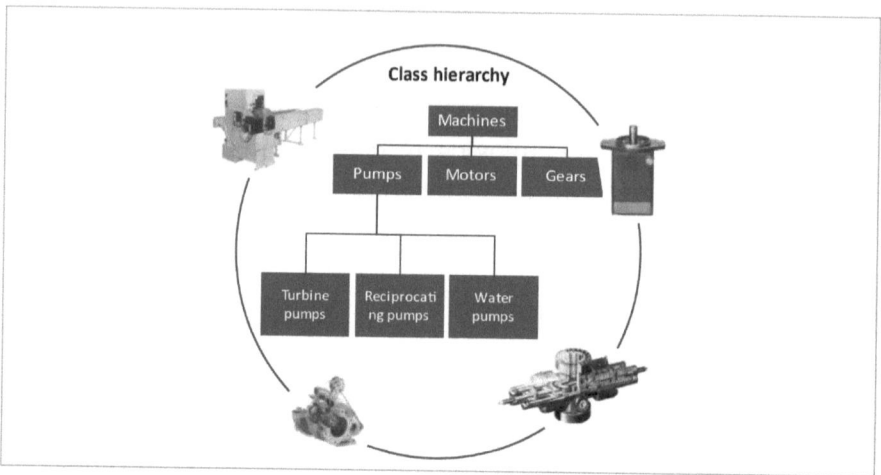

Figure 3-18. *Classification system*

Characteristics can be valued manually or automatically. Depending on the control settings for a characteristic, values can be selected from a predefined list or entered freely. In the case of automatic valuation, a characteristic may refer to a table field or another characteristic.

3.15. Standard Fiori Apps for Equipment Master

There are several Fiori apps available for managing the equipment master in SAP EAM. The following are some examples.

- Create Equipment (App ID: F0702A): This app allows you to create new equipment master records with relevant details and attributes.

- Change Equipment (App ID: F0731A): With this app, you can modify and update existing equipment master records by making changes to various fields and properties.

- Display Equipment (App ID: F0732A): This app
  provides a read-only view of equipment master
  records, allowing you to view detailed information and
  attributes of equipment.

- Equipment List (App ID: F0721): This app enables you
  to search and display a list of equipment based on
  specific criteria, providing a comprehensive overview
  of equipment records.

- Equipment Analysis (App ID: F0740A): This
  app offers analytical capabilities to analyze and
  evaluate equipment data, such as equipment status,
  maintenance history, and performance indicators.

- Equipment Where-Used List (App ID: F0735A):
  With this app, you can determine where a particular
  equipment is used by displaying its relationships and
  associations with other objects, such as functional
  locations or orders.

Please note that the availability and specific functionality of Fiori apps may depend on your SAP EAM system version, configuration, and licensing.

Figure 3-19 shows the Technical Object window in Fiori apps.

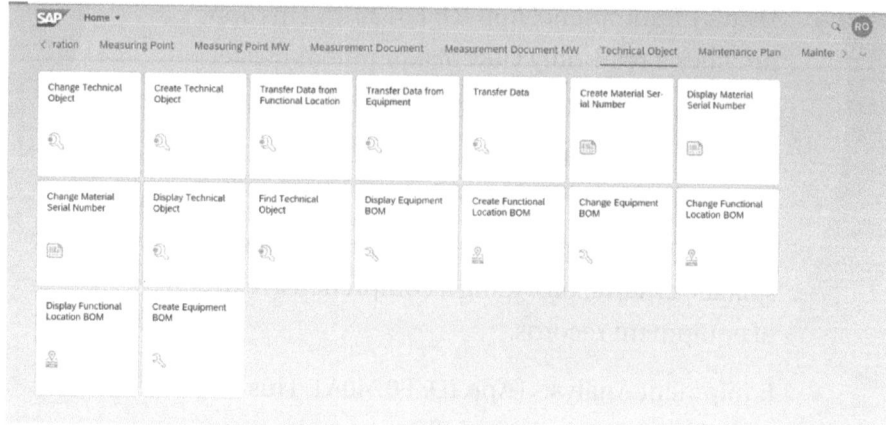

Figure 3-19. *Technical object Fiori apps*

3.16. Customizing Settings for Equipment Master

This section describes the Customizing settings for the equipment master (see Figure 3-20 and Table 3-2).

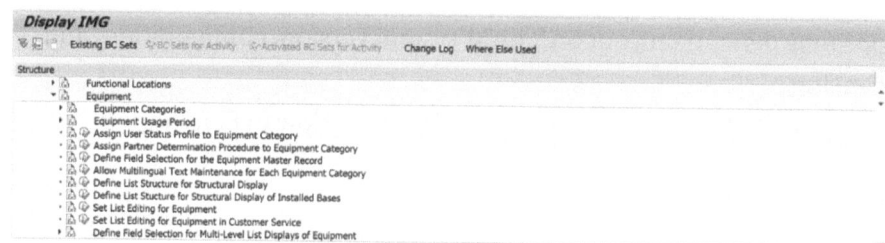

Figure 3-20. *IMG settings for equipment master*

Table 3-2. *IMG Settings for Equipment*

Field Name and Data Type	Menu Path
Define User Status Profile	Plant Maintenance and Customer Service → Master Data in Plant Maintenance and Customer Service → Basic Settings → Maintain User Status
Assign User Status Profile	For functional locations: Plant Maintenance and Customer Service → Master Data in Plant Maintenance and Customer Service → Technical Objects → Functional Locations → Define Category of Functional Location For equipment: Plant Maintenance and Customer Service → Master Data in Plant Maintenance and Customer Service → Technical Objects → Equipment → Equipment Categories → Maintain Equipment Category
Automatic Creation of Equipment	Financial Accounting → Asset Accounting → Master Data → Automatic Creation of Equipment Master Records
Business Views	Plant Maintenance and Customer Service → Master Data in Plant Maintenance and Customer Service → Technical Objects → Equipment → Equipment Categories → Define Additional Business Views for Equipment Categories
Installation at Functional Location	Plant Maintenance and Customer Service → Master Data in Plant Maintenance and Customer Service → Technical Objects → Equipment → Equipment Usage Period → Define Installation at Functional Location

(*continued*)

Table 3-2. (*continued*)

Field Name and Data Type	Menu Path
Usage History	Plant Maintenance and Customer Service → Master Data in Plant Maintenance and Customer Service → Technical Objects → Equipment → Equipment Usage Period
Equipment Category	Plant Maintenance and Customer Service → Master Data in Plant Maintenance and Customer Service → Technical Objects → Equipment → Equipment Categories → Maintain Equipment Category
Field selection for Equipment Master record	Plant Maintenance and Customer Service → Master Data in Plant Maintenance and Customer Service → Technical Objects → Equipment → Define Field Selection for the Equipment Master Record
Warranty Types	Plant Maintenance and Customer Service → Master Data in Plant Maintenance and Customer Service → Basic Settings → Warranties → Check Warranty Types
Warranty Categories	Plant Maintenance and Customer Service → Master Data in Plant Maintenance and Customer Service → Basic Settings → Warranties → Check Warranty Categories
Number Range	Plant Maintenance and Customer Service → Master Data in Plant Maintenance and Customer Service → Technical Objects → Equipment → Equipment Categories → Define Number Ranges
Object Type	Plant Maintenance and Customer Service → Master Data in Plant Maintenance and Customer Service → Technical Objects → General Data → Define Types of Technical Object

(*continued*)

Table 3-2. (*continued*)

Field Name and Data Type	Menu Path
Partner Functions	Plant Maintenance and Customer Service → Master Data in Plant Maintenance and Customer Service → Basic Settings → Partners → Copy Partner Functions to Master and Movement Data
Partner Determination Procedure	Plant Maintenance and Customer Service → Master Data in Plant Maintenance and Customer Service → Basic Settings → Partners → Define Partner Determination Procedure and Partner Function
View Profile	Plant Maintenance and Customer Service → Master Data in Plant Maintenance and Customer Service → Technical Objects → General Data → Set View Profile for Technical Objects

3.17. Summary

In this chapter, you learned how SAP equipment master provides a comprehensive understanding of managing equipment data within SAP. It covers the creation, modification, and deletion of equipment records, allowing users to maintain accurate and up-to-date information. The chapter emphasized the importance of equipment classification and hierarchy, enabling efficient organization and easy retrieval of equipment data. It also explores the integration of equipment master data with other SAP modules, such as maintenance and procurement, to streamline business processes. The chapter discussed various functionalities, including equipment status management, serial number handling, and attachments management. Overall, this chapter equips users with the knowledge and tools to effectively manage equipment master data and optimize equipment-related operations within the SAP environment.

3.18. Linear Asset Management

A project team is evaluating the functionalities of linear asset management to facilitate the mapping of technical assets with a linear structure and enable referencing of linear data during maintenance processing, as part of the optimization of business processes in your SAP ERP system.

In SAP EAM (Enterprise Asset Management), linear asset management refers to the management of assets that have a linear or linear-like structure, such as roadways, railway tracks, pipelines, and power transmission lines. These assets are typically extensive and have distinct characteristics compared to traditional fixed assets.

Linear asset management in SAP EAM involves the integration of spatial data (geographical information) with asset management processes to enable efficient planning, monitoring, and maintenance of linear assets throughout their lifecycle.

The following are some key features and functionalities of linear asset management in SAP EAM.

- **Linear asset hierarchy**: SAP EAM allows you to define a hierarchical structure for linear assets, including main lines, sections, and sub-sections. This structure helps in organizing and managing the assets effectively.

- **Linear asset master data**: You can create and maintain master data for linear assets, including details like start and end points, distances, coordinates, and associated attributes. This information helps in accurately locating and identifying the assets.

- **Linear asset networks**: SAP EAM enables you to define networks or routes for linear assets, representing their physical connections or relationships. This network concept helps in visualizing the asset layout and optimizing maintenance activities.

- **Linear asset planning**: You can plan and schedule maintenance activities for linear assets based on predefined criteria, such as distance, time, or condition. The system can generate work orders, schedule inspections, and assign resources accordingly.

- **Linear asset inspections**: SAP EAM allows you to perform inspections and condition monitoring of linear assets using various methods, such as mobile devices or remote sensing technologies. The data collected during inspections helps in assessing asset health and identifying potential issues.

- **Linear asset work management**: The system supports work management processes specific to linear assets, such as work order management, resource allocation, and tracking of work progress along the asset's length.

- **Linear asset analytics**: SAP EAM provides analytical capabilities to generate reports and perform data analysis on linear assets. This helps in making informed decisions, optimizing maintenance strategies, and improving asset performance.

By leveraging the features of linear asset management in SAP EAM, organizations can enhance the efficiency, reliability, and safety of their linear assets while reducing maintenance costs and downtime.

3.19. Structuring Linear Assets

Linear assets refer to technical systems with a linear infrastructure, and their conditions and properties can vary across different sections. To display, manage, and maintain these assets effectively, you can input the following details as linear data.

- Start and end points

- Length and measurement unit (e.g., kilometers or miles)

- Marker information (including start and end points, distance from a marker to a start point, distance from a marker to an end point) and its measurement unit

- Offset information (e.g., horizontal or vertical offset) and its measurement unit

Some examples of linear assets are pipelines, roads, railways, overhead power lines, and cables.

Within the system, you can create linear assets as technical objects, including functional locations and equipment, and store associated linear data. Furthermore, you can perform maintenance tasks for these technical objects, which generate notifications, maintenance orders, measurement documents, and other relevant items. These maintenance tasks offer several benefits, allowing you to do the following.

- Effectively manage the condition of your linear assets

- Identify areas of damage or defects using information such as the start point, end point, and offset

- Handle various types of maintenance tasks, including planned, unplanned, and preventive activities

3.20. Linear Reference Pattern

Linear asset elements are assigned a linear reference pattern (LRP) to serve as reference points for describing events or locations. LRPs are utilized to indicate the distance from the start point to the specific location where a marker is positioned. Figure 3-21 shows LRP in the Fiori app.

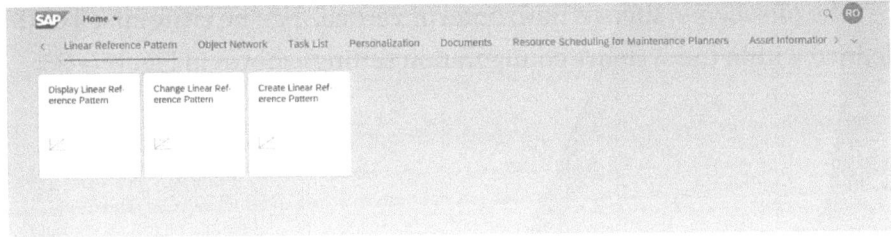

Figure 3-21. *Manage linear reference pattern*

3.20.1. Linear Data Offset

Offset refers to the distance or displacement of an object in relation to the reference line of a linear technical asset. For instance, it could represent the distance of a single mast from the center of a track or the distance of a manhole cover from the road. Offsets are employed to accurately describe the precise position of a technical object shown in Figure 3-22.

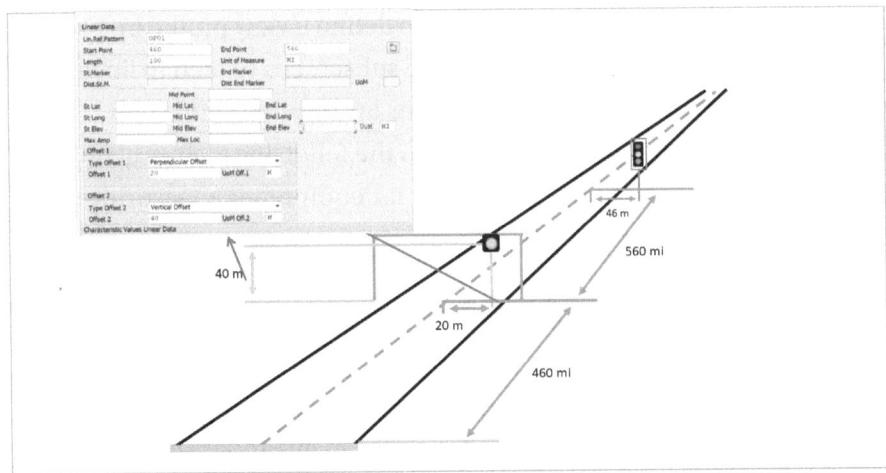

Figure 3-22. *Linear data offset*

The offset type, such as horizontal or vertical, can be customized and defined within the system's configuration settings shown in Figure 3-23.

| Offset Types | | | | |
OTC	Offset Type Desc.		UoM	Default Offset
LO	Longintudinal Offset		FT	Not a Default for Any Of... ▾
PO	Perpendicular Offset		FT	Default for Offset 1 ▾
VO	Vertical Offset		FT	Default for Offset 2 ▾

Figure 3-23. *IMG settings for offset types*

3.20.2. Linear Technical Objects

To configure the system, it is necessary to define a view profile in Customizing. This profile should include subscreens related to linear asset management for both functional locations and equipment.

Linear data fields have been added to the following technical objects as part of enhancements: functional locations, equipment, measuring points.

3.21. Linear Data in Maintenance Processing

The following objects have been enhanced with linear data fields for maintenance processing.

- Maintenance notification

- Work order header

- Work order operation

- Work order confirmations

- Measurement documents

- Maintenance plan item

- Relevant list reports

3.22. Network Object Link and Network Attributes

Linear data can be entered for both object networks and individual object links. Streamlining maintenance processes is possible by utilizing LRP markers assigned at the network level.

During object network maintenance, the following features can be employed.

- Managing object links

- Generating notifications and orders

- Accessing various lists such as notification lists, scheduling overviews, and multilevel lists

- Utilizing a graphical view of object networks, which can be customized according to specific customer requirements. You can manage object networks in Fiori apps (see Figure 3-24).

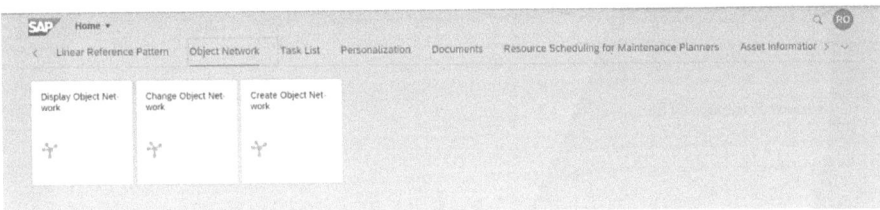

Figure 3-24. *Manage object network*

3.22.1. Assigning Network Attributes

You can enter additional information for object networks through attribute values and attribute properties via the running transaction, as shown in Figure 3-25.

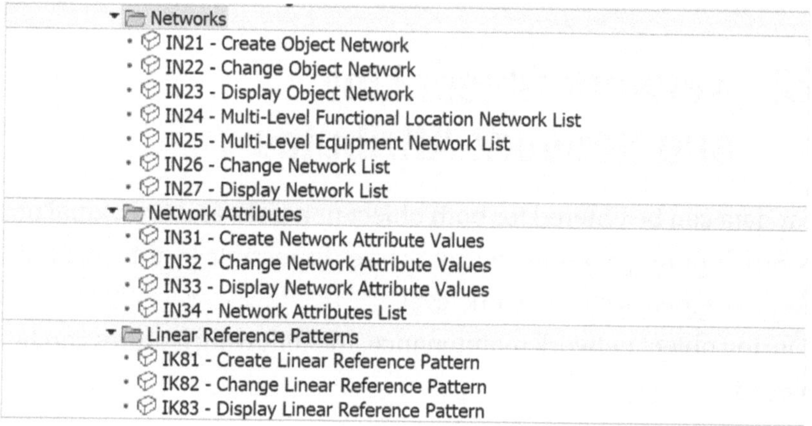

```
▼ 🗁 Networks
   • ⬡ IN21 - Create Object Network
   • ⬡ IN22 - Change Object Network
   • ⬡ IN23 - Display Object Network
   • ⬡ IN24 - Multi-Level Functional Location Network List
   • ⬡ IN25 - Multi-Level Equipment Network List
   • ⬡ IN26 - Change Network List
   • ⬡ IN27 - Display Network List
▼ 🗁 Network Attributes
   • ⬡ IN31 - Create Network Attribute Values
   • ⬡ IN32 - Change Network Attribute Values
   • ⬡ IN33 - Display Network Attribute Values
   • ⬡ IN34 - Network Attributes List
▼ 🗁 Linear Reference Patterns
   • ⬡ IK81 - Create Linear Reference Pattern
   • ⬡ IK82 - Change Linear Reference Pattern
   • ⬡ IK83 - Display Linear Reference Pattern
```

Figure 3-25. *Transaction code of linear asset master data*

Note   During the customizing process, it is required to define network attribute categories. These categories encompass network attributes, network attribute properties, permissible property values, and the association of attribute properties with attributes. To ensure relevance to specific object networks, a network attribute category is assigned to a network type.

3.23. Customizing Settings for Linear Asset Master Data

This section describes the Customizing settings for Linear Asset Master Data (see Figure 3-26 and Table 3-3).

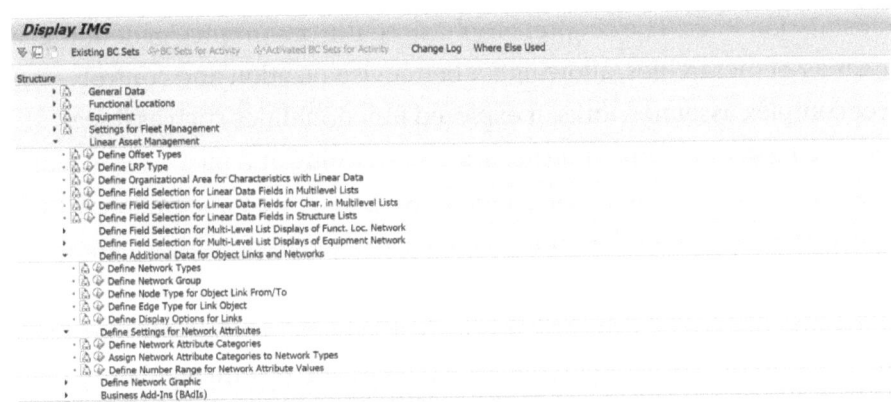

Figure 3-26. *IMG settings for linear asset*

Table 3-3. *IMG Settings for Linear Asset Management*

Field Name or Data Type	Menu Path
Linear asset management	Plant Maintenance and Customer Service → Master Data in Plant Maintenance and Customer Service → Technical Objects → Linear Asset Management

3.24. Summary

In this chapter you learned how linear assets provides a comprehensive overview of managing linear assets in SAP. It covered the creation, maintenance, and monitoring of linear asset master records, enabling effective management of assets such as roads, railways, pipelines, and transmission lines. The chapter emphasized the importance of defining linear asset hierarchies, allowing for better visualization and control over complex asset networks. It explored functionalities such as linear referencing systems, enabling accurate positioning and measurement of assets along linear routes. The chapter also highlighted integration with other SAP modules, such as maintenance and geographic information systems (GIS), to streamline asset management processes. Overall, this chapter equips users with the knowledge and tools to efficiently manage linear assets, optimize maintenance activities, and ensure regulatory compliance within the SAP EAM environment.

3.25. Bill of Materials

In SAP, a bill of materials (BOM) is a structured list that defines the components, subassemblies, and raw materials required to produce a finished product. It represents the hierarchical structure of a product and captures the relationships between its various parts.

The BOM in SAP serves as a central repository of information for manufacturing processes and is utilized across multiple modules, including Production Planning (PP), Materials Management (MM), and Sales and Distribution (SD).

A BOM typically includes the following information.

- **Header**: Contains general details such as the BOM number, description, and validity dates.

- **Items**: Each item represents a component or assembly required for the production process. It includes information such as material number, quantity, unit of measure, and item category.

- **Relationships**: BOMs can have various types of relationships, such as parent-child relationships between assemblies and components, alternative components, and phantom assemblies (used for planning purposes without physical existence).

- **Routing**: BOMs can be associated with routing data, which defines the sequence of operations and work centers required for manufacturing.

BOMs are crucial for planning production, material requirements, and cost estimation. They provide a comprehensive view of the components needed to manufacture a product, enabling accurate production planning, procurement, and costing calculations within SAP.

3.26. Structuring a Bill of Materials

A bill of materials (BOM) serves various purposes within a company, and the specific use of a BOM can vary based on the company's area of operation.

Let's look at examples of BOM uses.

- **Engineering/Design BOM** encompasses all the components of a product from an engineering or design perspective, including their technical specifications. It is generally not specific to a particular order.

- **Production BOM** focuses on items relevant to production and assembly status. For instance, it may include only production-related items with process-oriented data required for assembly.

- **Costing BOM** represents the product structure and serves as the basis for automatically determining material usage costs. It excludes items that are not relevant for costing purposes.

- **Maintenance BOM**, unlike other types of BOMs, contains only items essential for maintenance activities. It serves two primary functions.

  - **Structuring the object**: This provides a clear maintenance perspective on its structure.

  - **Spare part planning in orders**: A maintenance BOM facilitates planning spare parts when creating a maintenance order.

3.26.1. BOM Categories

The following describes BOM categories.

- **Material BOM** is directly linked to a material master record, which includes descriptive and control data such as measurements, weight, material category, and industry. It consists of the individual components (materials or assemblies) required for the object.

- **Equipment or Functional Location BOM** is utilized to describe the structure of an equipment or functional location and assign spare parts specifically

for maintenance purposes. It helps in organizing
and planning maintenance activities related to the
equipment or functional location.

- **Creation of BOM** provides the option to create BOMs
  of all categories, either with or without utilizing a
  template. When using a template, the following
  guidelines apply to each category shown in Figure 3-27.

  - To create a functional location BOM or equipment
    BOM, you can use material BOMs, equipment
    BOMs, and functional location BOMs as templates.

  - To create a material BOM, you can only use
    material BOMs as templates.

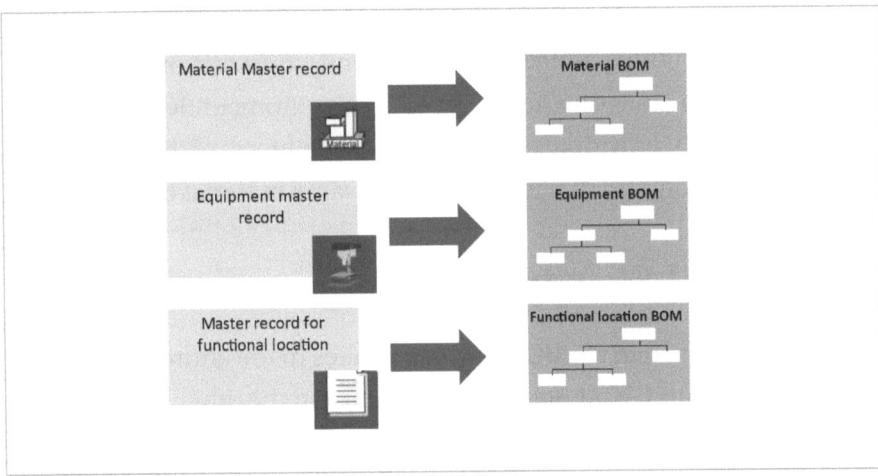

Figure 3-27. *BOM categories*

3.26.2. Assigning a BOM

BOMs can be linked to technical objects in two ways: through direct assignment and indirect assignment.

- For direct assignment, a BOM is directly created for the specific object, be it a functional location or equipment.

- For indirect assignment, a BOM is assigned to the technical object's master record using a material master record. This is accomplished by entering the material number in the Construction Type field within the technical object's master record. In this case, the BOM is linked to the material rather than the individual technical object. Indirect assignment is recommended when multiple objects share the same construction type, as there's no need to create an equipment BOM for each individual object. Instead, all objects of the same construction type utilize the same material BOM.

3.26.3. BOM Structure

The data maintained in the BOM header applies to the entire BOM and its components (BOM items or subitems). On the other hand, the components of a technical object are represented by the BOM items. The item data is specific to each individual item within the BOM and is not valid for the entire BOM.

3.26.4. Usage of BOM in Order

For each operation in the maintenance order, you have the option to schedule BOM components from the reference object (materials) that are necessary to carry out the task.

If the scheduled materials for the maintenance order are available in stock, they are reserved in the warehouse. The automatic creation of the reservation is determined by item category S, which represents stock materials.

However, if you schedule non-stock materials, the process flow differs. In such cases, the scheduling triggers the initiation of ordering using item category N, which leads to the automatic creation of a purchase requisition.

3.26.5. Item Status in BOM

Item status is defined based on the BOM usage. The item status encompasses various indicators related to engineering/design, production, costing, and more.

These item status indicators play a crucial role in controlling the process flow in subsequent work areas and the selection of items during BOM explosion. They determine whether certain processing is required, permitted, or excluded in these areas. If processing is supported in a specific work area, you can maintain data specific to that area for the items. For example, only items relevant to production are copied to the production order.

To manage structure elements or items relevant to maintenance in a functional location BOM or equipment BOM, you must choose a BOM usage that supports maintenance-related items. Maintenance-related BOM maintenance is performed separately for different areas, such as engineering/design and production, each with its own usage. If multiple BOMs with different uses are created for the same material, the system stores each BOM with a unique internal number corresponding to its usage.

Assemblies are structural components within a BOM, bringing together spare parts pertinent to maintenance activities.

3.26.6. Maintenance Assemblies

Maintenance assemblies are materials categorized as material type IBAU. The material master record for maintenance assemblies only contains essential and classification data, which means that inventory management is not available for these assemblies.

Material Assembly in SAP EAM involves grouping multiple materials or components as a single entity for maintenance or construction purposes. It streamlines processes and stock management by treating the assembly as a single item. In contrast, Material refers to individual components or items that are tracked and managed separately in the system.

For the following purposes, you can utilize maintenance assemblies as structural components within a BOM.

- Grouping similar materials under a single node.

- Tracking costs in the PMIS without requiring inventory management.

However, you also have the flexibility to select materials of other material types as maintenance assemblies if needed.

The Maintenance Assembly Indicator is activated in the Status/Long Text section of the BOM items screen and can only be assigned to items relevant for maintenance. When performing maintenance tasks, items designated as maintenance assemblies appear as structural elements within the operational system. They provide a means to provide more detailed descriptions of items within the operational system, such as specifying the potential location of damage shown in Figure 3-28.

The diagram shown in Figure 3-28 illustrates a hierarchical BOM for an electric pump. The BOM includes the following components.

- Stock item (L) - Motor

- PM structure element (I) - Shaft assembly

- PM structure element (I) - Control electronics

Each item has its own BOM, which includes the parts relevant for maintenance.

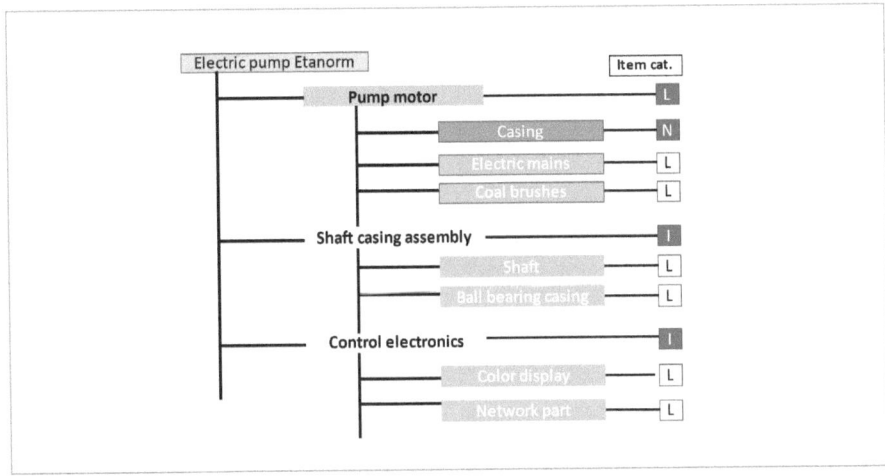

Figure 3-28. *Multilevel BOM with different categories*

The overall BOM indicates that the complete motor is stocked as a spare part. Coal brushes and electric mains are also kept in inventory as spare parts. However, the casing is procured externally and not managed as part of the inventory.

The shaft casing assembly and control electronics are not stocked as complete parts (see Figure 3-29). Instead, they are managed in the inventory and can be reserved for maintenance tasks.

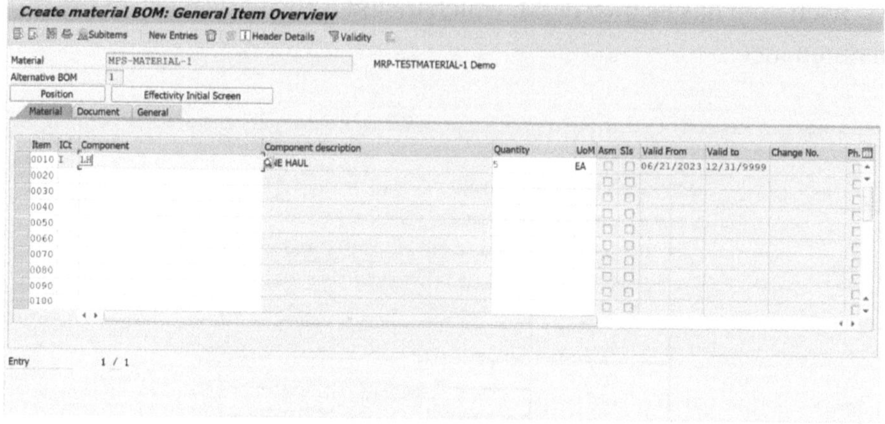

Figure 3-29. *BOM master data*

A BOM can handle essential data for plant maintenance at the technical object location, ensuring spatial validity within the plant. This includes organizing workstations where all necessary preparatory tasks, such as material planning and task list creation, are carried out. In such cases, a plant-specific BOM can be created.

Several checks are performed, such as ensuring a material master record with plant data exists for the relevant plant in the BOM header. The system verifies whether plant data is available for the material components when adding items. If the checks pass, the system copies the material into the material BOM.

Alternatively, generating a BOM known as a group BOM does not require a specific plant reference is possible. This type of BOM is typically utilized when an engineer creates a BOM during the design phase, which can later be assigned to one or multiple plants for production. In this case, the system solely verifies the availability of material master records and does not conduct plant-specific material checks.

3.27. Standard Fiori Apps for Bill of Materials

Several SAP Fiori apps are available for managing SAP BOM within SAP technologies. The following are some of these apps.

- **Manage Bill of Materials** (App ID: F1708): This app allows users to create, edit, and view BOMs. It provides a user-friendly interface for managing BOM data, including material components, quantities, and dependencies.

- **Display Material BOM** (App ID: F1462): This app lets users view detailed information about a specific material BOM, including its structure, components, and associated data. It provides an intuitive interface for navigating and exploring BOM details.

- **Change Material BOM** (App ID: F1464): With this app, users can change existing material BOMs, such as adding or removing components, updating quantities, or modifying other BOM attributes. It simplifies the process of maintaining and updating BOM data.

- **Where-Used List for Material BOM** (App ID: F1592): This app allows users to find all the places where a specific material BOM is used within the SAP solution. It provides visibility into the various production processes or assemblies that rely on the selected BOM.

- **Display Plant-Specific BOM** (App ID: F1709): This app allows users to view a BOM specific to a particular plant. It provides plant-specific information, including plant-specific components, alternate BOMs, and other plant-related details.

These are just a few examples of the SAP Fiori apps for managing SAP BOM. The availability and specific functionalities of these apps may vary depending on the version and configuration of the SAP software solution (see Figure 3-30).

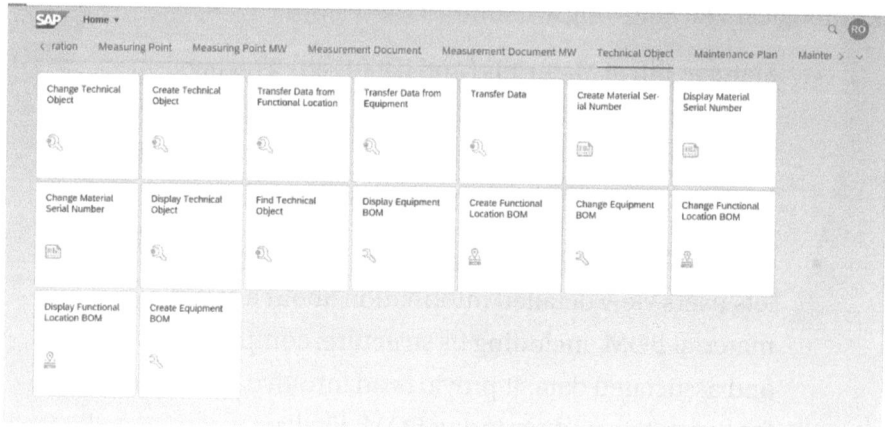

Figure 3-30. *Manage bill of material*

3.28. Customizing Settings for Bill of Materials

This section describes the Customizing settings for the bill of materials (see Figure 3-31 and Table 3-4).

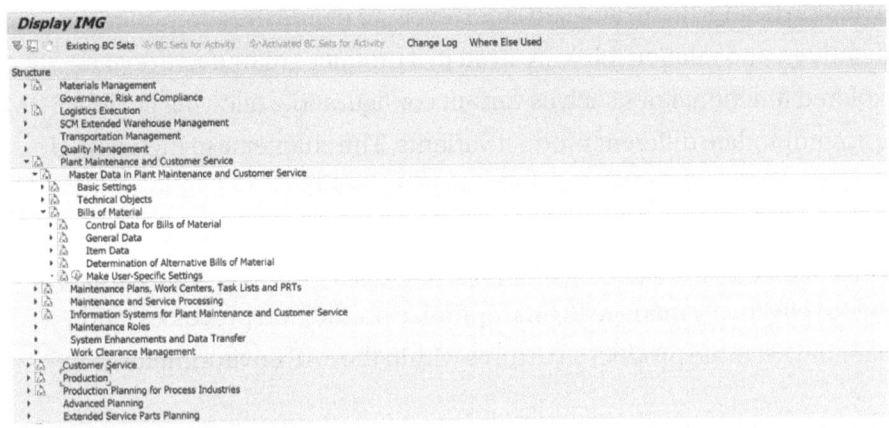

Figure 3-31. *IMG Settings for BOM*

Table 3-4. *IMG Settings for Bill of Materials*

Field Name or Data Type	Menu Path
Item Categories	Plant Maintenance and Customer Service → Master Data in Plant Maintenance and Customer Service → Bill of Materials → Item Data → Define Item Categories
BOM Usage	Plant Maintenance and Customer Service → Master Data in Plant Maintenance and Customer Service → Bill of Materials → General Data → BOM Usage → Define BOM Usages

3.29. Summary

This chapter explained how SAP BOM provides a comprehensive understanding of managing BOM data within SAP. It covered creating, maintaining, and using BOMs, enabling efficient production and assembly processes. The chapter also emphasized the importance of BOM structure, including header, item, and sub-item levels, to define the components, quantities, and relationships involved in a product's manufacturing. It explored functionalities such as variant configuration, allowing flexible BOMs to accommodate different product variants. The chapter also highlighted integration with other SAP modules, such as production planning and procurement, to streamline the supply chain and ensure accurate material requirements planning. This chapter equips users with the knowledge and tools to effectively manage BOMs, optimize production processes, and maintain accurate product structures within the SAP environment.

3.30. Serial Numbers

Serial numbers in SAP are used to uniquely identify individual items or products within the system. They provide a way to track and manage individual units throughout their lifecycle, from production to sales and post-sales activities. The following are some common uses of serial numbers in SAP.

- **Traceability and warranty management**: Serial numbers enable accurate tracking of products from manufacturing to delivery and beyond. They help identify the specific unit's history, components used, production date, and other relevant information. This traceability is crucial for warranty management, product recalls, quality control, and addressing customer inquiries.

- **Inventory management**: Serial numbers allow for precise inventory management by tracking the movement and location of individual items. Each serial number can be associated with specific warehouses, storage bins, or sales orders, providing granular visibility into stock levels, movements, and valuation.

- **Sales and distribution**: Serial numbers are commonly used in sales and distribution processes to manage serialized product sales. They ensure accurate identification of products sold, enable tracking of sold items to specific customers, and facilitate after-sales support such as servicing, repairs, or returns.

- **Service management**: Serial numbers are vital in managing service and maintenance activities. They enable service technicians to identify the exact unit requiring maintenance, track service history, record repairs and replacements performed, and manage warranty claims or service contracts.

- **Product lifecycle management**: Serial numbers support the management of the entire product lifecycle, from production to disposal. They enable detailed analysis of product performance, customer feedback, maintenance costs, and other metrics to drive improvements in product design, production processes, and customer satisfaction.

By leveraging serial numbers in SAP, businesses can achieve better inventory control, streamlined processes, improved customer service, and enhanced visibility into product-related data for decision-making and compliance purposes.

3.30.1. Working with Serial Numbers in SAP EAM

To initiate the process of dismantling and transferring a faulty piece of equipment to the warehouse, the equipment must have serialization. Serialization activates the tab for serialization data and establishes a link between the equipment and a specific serial number associated with a material. To refurbish the defective equipment, which has serialization, the serial number establishes a connection with a material that undergoes condition-based evaluation. The material is divided into distinct partial stocks based on conditions, such as "as new" (C1), "refurbished" (C2), and "faulty" (C3).

Once the equipment has been refurbished, it is returned to the warehouse using the corresponding linked material and serial numbers.

Through a single step, it is possible to remove equipment from one storage location and install it at another equipment or functional location. Serialization facilitates the management of equipment inventory by enabling efficient tracking and control of individual pieces of equipment.

The following are sales/customer service examples.

- In sales and customer service, a common requirement is to deliver a serialized material item to a customer while ensuring its unique identification. The material stock can be stored in the warehouse collectively without tracking individual items.

- By associating the material with a predefined serial number profile, serialization is automatically applied during the good's issue. This process assigns serial numbers to each piece. If the serial number profile includes equipment requirements, an equipment master is automatically created with the assigned serial number.

- You can store location or main work center data based on the equipment. In the event of a customer complaint, it is possible to record a service message or service order specific to the respective piece of equipment.

3.30.2. **Material and Serial Number**

The serial number profile can be inputted in the following material master views shown in Figure 3-32.

- Sales: General/plant data

- General plant data/storage 2

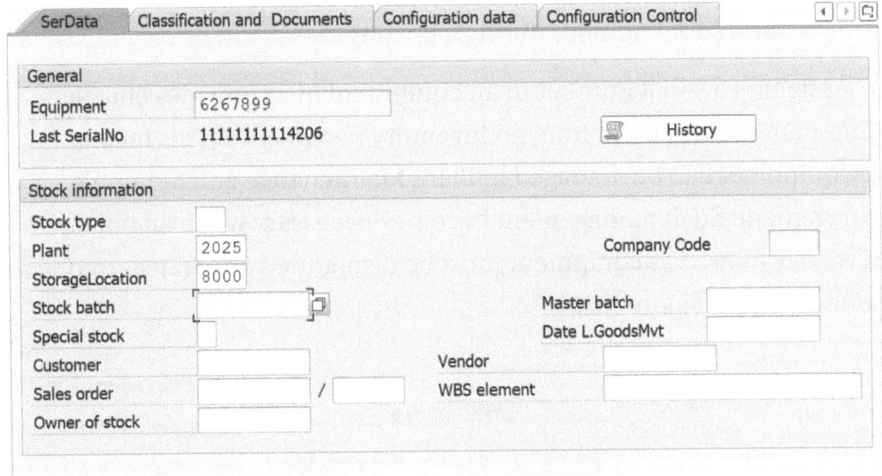

Figure 3-32. *Serial number master data*

When the assignment is made in the Sales: General/plant data view, the profile automatically appears in the General plant data/storage 2 view (and vice versa). It's important to note that the material and serial number combination is unique at the client level.

3.30.3. Equipment and Serial Number

Two choices exist for connecting the equipment, material, and serial number.

- The material and serial number can be manually assigned to existing equipment.

- The system can automatically generate the equipment and serial number during a posting.

Assigning a serial number to an equipment item provides effective equipment management from an inventory perspective. This means the equipment can be managed in Plant Maintenance and Materials Management. Such management becomes necessary when an object previously treated as equipment must be dismantled and transferred to the warehouse (see Figure 3-33).

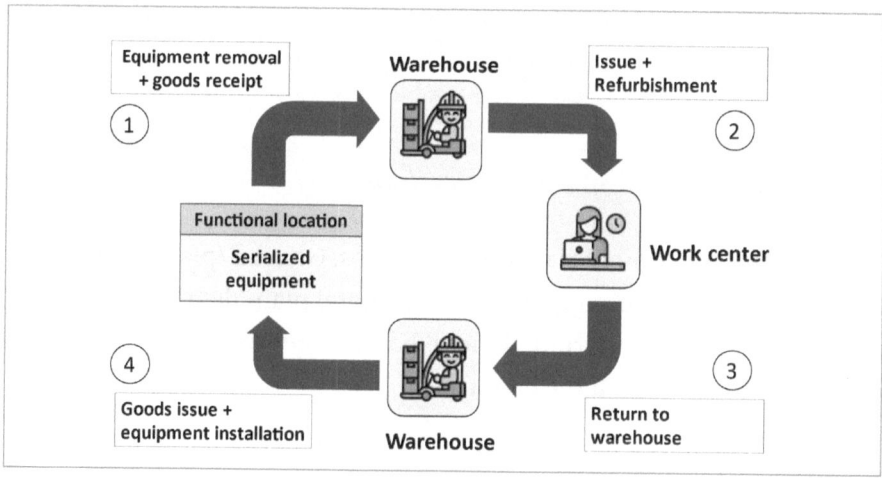

Figure 3-33. *Example serialization process in EAM*

The serial number is assigned in the serialization data section of the equipment master record. If a material already has a serial number profile for individual equipment pieces within the equipment master record, the system displays the last assigned serial number for that material. This allows a direct connection between the new serial number and the previously assigned one.

The relationship between the material and equipment can be managed in the following ways.

- Synchronizing the equipment number and serial number with each other.

- Synchronizing the material associated with the equipment with the material specified in the Construction Type field of the equipment (Structure tab).

Note   The display of the Serial Data view in the equipment master record follows the presetting in the Customizing settings for the equipment category. If the view is not initially activated in Customizing, it can be displayed later.

3.30.4. Serial Number Profile

A four-digit key defines a serial number profile and contains data that specifies the conditions and operations for assigning serial numbers to material items. These profiles are set up during system configuration in the Plant Maintenance section of Customizing. Parameters that can be defined include the business transactions requiring checks, existence requirements for serial numbers, automatic creation of serial numbers, existence requirements for equipment pieces, and equipment category.

3.30.5. Stock Check with Serial Number

A *stock check* refers to verifying stock information in the serial number record during goods movement.

During the stock check, the system compares the stock information in the serial number with the stock data in materials management. The system follows the predefined setting to halt the posting if discrepancies are detected.

The master record for serial numbers contains stock information, including stock type, plant, and storage location; stock batch and master batch; special stock; and vendor and sales order.

Serial numbers are not only updated for the specific plant and storage location but also for all stock types (e.g., unrestricted-use stock, stock in quality inspection, blocked status returns, stock in transfer, stock in transit, blocked stock) and special stock (e.g., open order quantity, project stock, customer, and vendor consignment stock).

Serial numbers created before the R/3 4.5A Release do not automatically include stock information. However, this information is gradually added with each serial number movement.

3.30.6. Check Goods Movement

During goods movement, the relevant stock fields are examined and validated. The system compares this data with the inventory management records. Based on the serial number profile configuration, if there is a mismatch, the movement can be prohibited, allowed, or allowed with a warning. After the movement, the serial number is updated with the current stock data from inventory management.

To enable stock verification, it must be activated in the serial number profile through customizing.

There are specific programs available to assist in monitoring stock information.

- The RIMMSF00 program compares the stock data in the serial number master data with the stock data in inventory management. It verifies the cumulative serial number stock against each item of special stock in inventory management.

- The RISERNR9 program transfers the stock verification indicator from the serial number profile to individual serial numbers. It is important to execute this program if changes have been made to the indicator in the profile and goods movements have already been performed with the serial numbers.

You can manually modify stock information in the serial number under the following circumstances.

- When no plant information is provided

- When there is no active stock check

- When the stock check is active, but only with a warning

In cases where multiple users are concurrently trying to create serial numbers for the same material number, a lock mechanism is implemented to ensure consistent number assignment.

3.31. Standard Fiori Apps for Serial Numbers

Several Fiori apps are available for serial numbers in SAP, which provide serial number management and tracking functionalities. The following are some commonly used Fiori apps for serial numbers in SAP.

- Manage Serial Numbers (App ID: MMIM_
  SERIALNUMBERS): This app allows users to manage
  serial numbers for materials, including creating,
  editing, and displaying serial numbers. Users can also
  perform actions like goods receipt and issue, view serial
  number details, and perform stock inquiries.

- Create Serial Numbers (App ID: MMIM_
  SERIALNUMBERS_CREATE): This app enables users
  to create new serial numbers for materials. Users can
  enter relevant details such as material number, batch,
  equipment, and additional attributes to generate serial
  numbers.

- Serial Numbers Overview (App ID: MMIM_
  SERIALNUMBERS_OVERVIEW): This app provides
  a comprehensive overview of serial numbers for
  materials. Users can view and analyze serial number
  data, including stock levels, storage locations, and
  usage information. It also allows users to search and
  filter serial numbers based on various criteria.

- Serial Number Validation (App ID: MMIM_
  SERIALNUMBERS_VALIDATION): This app allows
  users to validate and verify the correctness of serial
  numbers. Users can check the uniqueness and integrity
  of serial numbers against predefined rules and criteria.

- Serial Number Stock Overview (App ID: MMIM_
  SERIALNUMBERS_STOCK): This app provides a
  detailed view of stock information for serial numbers.
  Users can check stock levels, movements, and locations
  associated with specific serial numbers.

These are just a few examples of the Fiori apps available for serial numbers in SAP. The availability and specific functionalities may vary based on the SAP software version and configuration.

3.32. Customizing Settings for Serial Numbers

This section describes the Customizing settings for the bill of materials (see Figure 3-34 and Table 3-5).

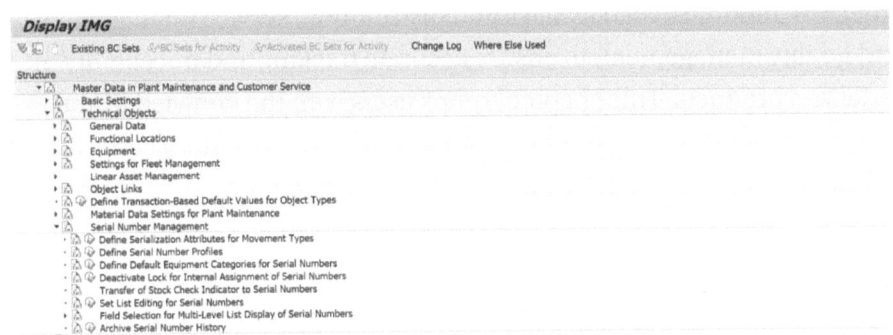

Figure 3-34. *IMG settings for serial number management*

Table 3-5. *IMG Settings for Serial Number*

Field Name or Data Type	Menu Path
Serial Number Profile	Plant Maintenance and Customer Service → Master Data in Plant Maintenance and Customer Service → Technical Objects → Serial Number Management → Define Serial Number Profiles
Serialization Attributes for movement types	Plant Maintenance and Customer Service → Master Data in Plant Maintenance and Customer Service → Technical Objects → Serial Number Management → Define Serialization Attributes for Movement Types

3.33. Summary

In this chapter, you learned how SAP Serial Numbers provides a comprehensive understanding of managing serial numbers within SAP. It covers creating, tracking, and managing serial numbers for individual items or products. The chapter emphasizes the importance of assigning unique serial numbers to track each item's lifecycle, warranty, and maintenance history. It explores functionalities such as serial number profile configuration, allowing customized settings based on specific business requirements. The chapter also highlights integration with other SAP modules, such as sales and distribution, service management, and quality management, to ensure accurate tracking and traceability of serialized products. This chapter equips users with the knowledge and tools to efficiently manage serial numbers, improve inventory control, and enhance customer service within the SAP environment.

3.34. Measuring Points and Counters

In SAP EAM, measuring points and counters are used to track and monitor various measurements and readings associated with equipment, functional locations, or other objects. They provide a means to capture and record data related to specific parameters or indicators, enabling organizations to monitor performance, conduct analysis, and make informed decisions. The following are the key uses of measuring points and counters in SAP.

- **Equipment maintenance**: Measuring points and counters are crucial for monitoring the condition and performance of equipment. Organizations can assess equipment health, identify potential issues, and schedule preventive maintenance or repairs by capturing temperature, pressure, vibration, or energy consumption measurements.

- **Quality control**: Measuring points and counters track quality-related measurements during production processes. For example, manufacturing can capture data such as dimensions, tolerances, or defect rates. This information helps ensure product quality, identify deviations, and take corrective actions.

- **Energy management**: Measuring points and counters are utilized for monitoring and management. Organizations can identify energy usage patterns, detect inefficiencies, and implement energy-saving initiatives by measuring energy consumption at various points within a facility or production process.

- **Performance analysis**: Measuring points and counters provide data for performance analysis and benchmarking. Organizations can measure and compare performance across different assets, locations, and time periods by tracking parameters such as production output, cycle times, or downtime. This analysis supports optimization efforts and helps identify areas for improvement.

- **Regulatory compliance**: Measuring points and counters assists in meeting regulatory requirements. Certain industries have specific monitoring and reporting obligations, such as environmental emissions or safety parameters. Measuring points and counters facilitate the collection of relevant data, ensuring compliance with regulations and enabling accurate reporting.

- **Predictive maintenance**: Organizations can develop predictive maintenance models by analyzing historical data from counters and measuring points. These models use the collected data to identify patterns, trends, or anomalies that indicate potential equipment failures. This proactive approach helps optimize maintenance planning, reduce unplanned downtime, and increase overall equipment effectiveness.

Overall, measuring points and counters in SAP provide a structured and centralized approach to capture and manage critical measurements and readings. They enable organizations to monitor, analyze, and optimize various aspects of their operations, leading to improved efficiency, productivity, and decision-making.

3.35. Working with Measuring Points and Counters

To enhance the monitoring of capacity utilization, your objects are equipped with measuring points and counters.

3.35.1. Measuring Points

Measuring points are specific physical or logical locations associated with technical objects where conditions are described, such as the temperature of the coolant in a nuclear power station or the rotations per minute of wind-driven power station blades. These measuring points are at the technical objects themselves (see Figure 3-35).

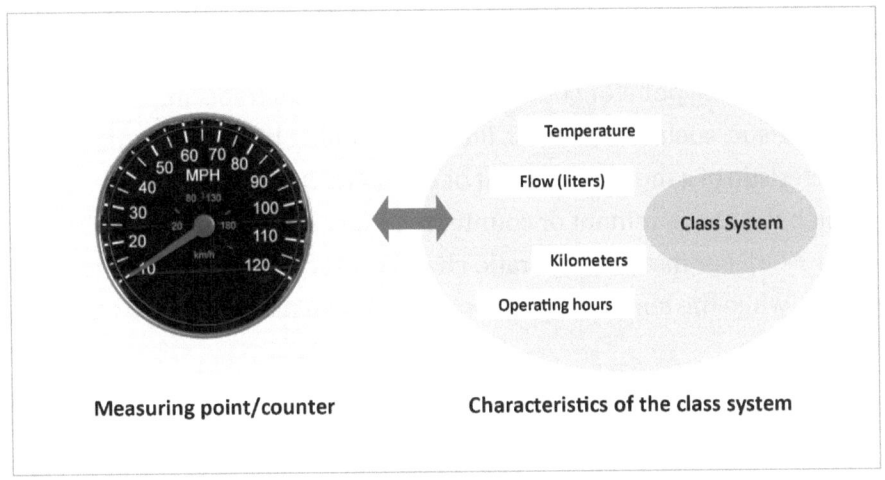

Figure 3-35. *Measuring points and characteristics*

3.35.2. Counters

Conversely, counters serve as resources that allow you to quantify the damage to an object or track the consumption or reduction of its useful life. Examples of counters include the mileage indicator of a motor vehicle or the electricity consumption meter of an electrically powered system. Like measuring points, counters are also located at the respective technical objects.

If you require maintenance for an object, you can input measurements or counter readings. These readings are essential for documenting the object's condition based on measurements or determining the maintenance schedule based on meter readings. In Customizing, you have the flexibility to define different types of measuring points.

3.35.3. Measuring Points and Characteristics

Every measuring point or counter is associated with a specific characteristic, such as kilometer, liter, or operating hours. The characteristic of a measuring point or counter determines the unit in which the measurement or counter readings are recorded (see Figure 3-36). For instance, the ratio characteristic may have the unit percent, while the temperature characteristic can have units like Celsius or Fahrenheit.

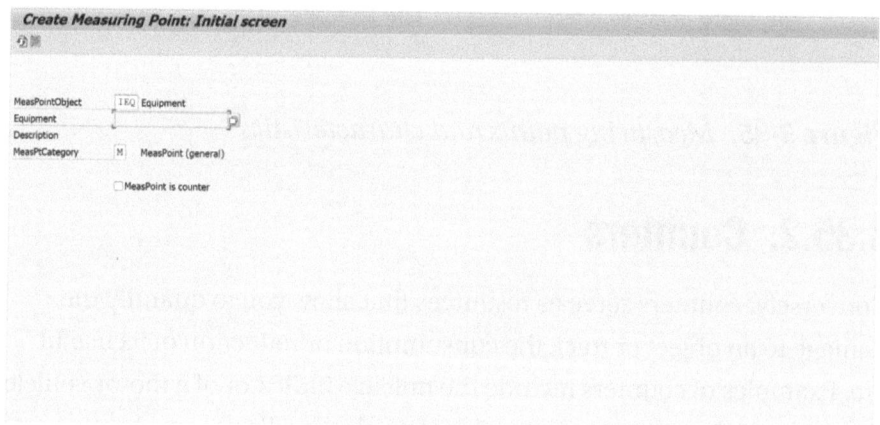

Figure 3-36. *Creating measuring point master data*

The classification system must create all the corresponding characteristics to ensure accurate maintenance of measuring points and counters for your objects. This system falls under the cross-application functions within SAP. Creating these characteristics is a crucial step before maintaining the measuring points and counters.

3.35.4. Measuring Documents

A measurement document is a comprehensive term that describes the data entered in the system following measurements taken at a measuring point or after a meter reading. The measurement document encompasses the following information (see Figure 3-37).

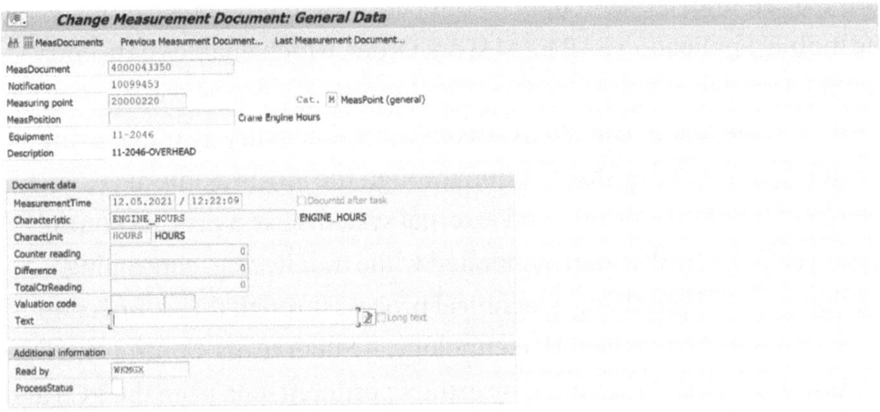

Figure 3-37. *Measuring document*

- **Measuring point data** includes details such as the measuring point number, measurement position, description, characteristic, and unit.

- **Measuring result data** encompasses information about the time of measurement/reading, the actual measurement/counter reading, and a qualitative assessment of the measurement result/counter reading.

- **Additional information** includes supplementary details like technician information, explanations, and more.

Moreover, assigning a processing status to the measurement document is possible. This status indicates if any action is required based on the measurement value or counter reading and determines whether existing tasks cover the need for action based on these readings.

The data can be manually entered into SAP technology through the SAP user interface (SAP GUI). Alternatively, measurement readings can be recorded from a maintenance notification or maintenance order using the mobile application SAP EAM Work Order, which enables data entry via mobile devices.

A barcode reader can aid in streamlining data entry by transferring the information to the SAP solution through the PM-PCS interface. This interface links SAP software and external systems like a process control system (PCS). The PCS data generated while monitoring, controlling, regulating, and optimizing a technical process through this connection can be transmitted to the SAP technology. A Supervisory Control and Data Acquisition (SCADA) system may extract pertinent data from the PCS and transmit it to the SAP technology.

Various customer exits are available to automate business processes, allowing for tasks such as automatically generating a malfunction report from a measurement document. These customer exits enable customization and automation of specific functionalities within SAP.

3.35.5. Measuring Point Entry List

For the simplified entry of routine counter and measurement readings across multiple objects, a master record known as a measurement reading entry list can be created. This list encompasses various measuring points or counters associated with different objects.

When determining measurement readings or taking counter readings, the corresponding values can be directly entered within this list. As before, the update occurs through a measurement document.

To transfer measurement and counter readings from external systems to the SAP solution, the PM-PCS interface can be utilized. This data is stored in measurement documents within the SAP technology and can be accessed by both the Plant Maintenance and Customer Service components.

Performance-based preventive maintenance allows maintenance planning based on counter readings managed for the relevant technical objects.

Measurement documents can be utilized to document subjects related to system security, work safety, and environmental protection.

The following are examples of external systems.

- Mobile data entry systems

- Process control systems

- Building control systems

- SCADA systems

Process control systems provide a wide range of processes, buildings, or infrastructure data. SCADA systems play a filtering role by extracting maintenance-relevant data, protecting the SAP technology from an excess of process data. Additionally, SCADA systems facilitate communication between one or more PCS and the SAP software solution.

3.36. Measurement Transfer and Counter Replacement

Measurement transfer in SAP EAM refers to the process of transferring measurement data, such as equipment readings or inspection results, from one location or system to another. It ensures accurate data continuity across different systems, enhancing maintenance and asset management activities.

Counter replacement involves updating counter values associated with equipment or assets during maintenance activities. It helps track usage and performance, aiding in predictive maintenance strategies by triggering alerts or actions based on predefined thresholds or intervals. This ensures optimal asset functionality and minimizes downtime.

3.36.1. Perform Measurement Reading Transfer

The following transfer scenarios are available.

- Transfer of measurements from one measuring point to another (1:1 relationship)

- Transfer of measurements from one measuring point to multiple others (1:n relationship)

- Transfer of counter readings from one counter to another (1:1 relationship)

- Transfer of counter readings from one counter to multiple others (1:n relationship)

The following conditions must be met for a successful transfer of measurement and counter readings.

- Both the measuring points and counter readings share the same characteristic.

- Both the measuring points or counters are part of the same object hierarchy. It is not possible to transfer measurement and counter readings across external object hierarchies.

In the case of a counter reading transfer, interval documents can be utilized. These documents aggregate the counter readings from subordinate levels within an asset structure, resulting in recurring entry documents. This approach significantly reduces the number of measurement documents. During a specific installation interval of an equipment, the same measurement document is used as long as the equipment remains installed at the same functional location or higher-level equipment.

If a measurement document is entered for a historical reference time, it is transferred based on the object structure that was valid during the reference time, even if there have been changes in the object structure.

When installing or dismantling a piece of equipment with a historical reference time, the system automatically cancels unnecessary measurement documents and creates new ones if required.

3.36.2. Counter Replacement

In case of a defective register or the need for an entire counter replacement, it is necessary to display the counter replacement in the system. This ensures that the discrepancy in the continuously increasing or decreasing readings between the old counter and the new counter is explained. Typically, the reading of the new counter differs from that of the old one.

The overall counter reading remains unchanged when entering a measurement document to document a counter replacement. Instead, the new reading of the newly installed counter should be entered with this measurement document at the beginning of its usage. This new reading does not impact the overall counter reading recorded in the previous measurement document.

3.37. Standard Fiori Apps for Measuring Points and Counters

The specific Fiori apps available for measuring points and counters in SAP may vary based on your SAP solution's specific version and configuration. However, let's look at general information about Fiori apps related to measuring points and counters in SAP.

- **Manage Measuring Points** (App ID: F2950): This app allows you to create, modify, and display measuring points. You can define measuring points to capture data related to equipment, functional locations, or other objects in your SAP system (see Figures 3-38 and 3-39).

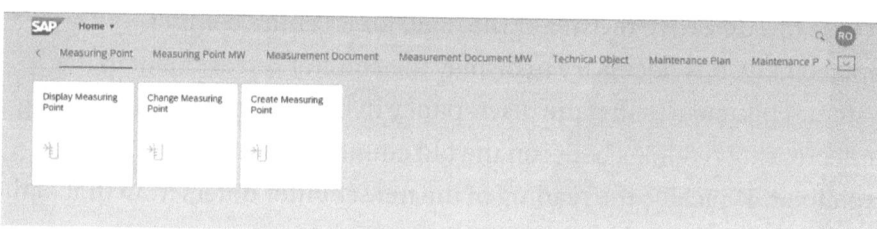

Figure 3-38. *Measuring point Fiori app*

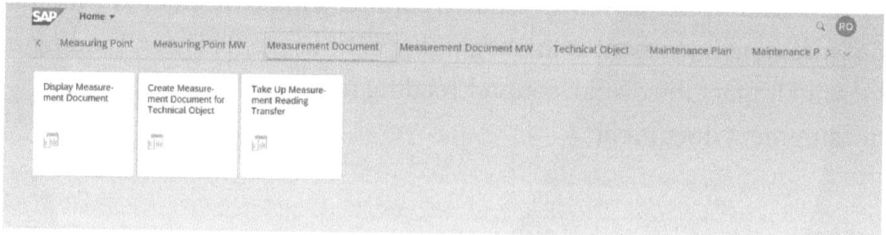

Figure 3-39. *Measuring document Fiori app*

- **Manage Counters** (App ID: F1908): This app enables you to create, modify, and display counters. Counters are used to track the usage or readings of measuring points over time. You can define counters based on specific measuring points and assign them to equipment or functional locations.

- **Monitor Measuring Points** (App ID: F1883): This app provides a dashboard-style overview of measuring points and their associated counters. You can view the current readings, historical data, and graphical representations of the measurements.

- **Monitor Counters** (App ID: F1909): This app allows you to monitor and analyze counter readings. You can view the counter values over time, compare readings, and generate charts or reports to analyze the data.

It's worth noting that SAP regularly updates and enhances its Fiori app portfolio, so there may be additional or updated apps available beyond the ones mentioned here. To get the most accurate and up-to-date information on the available Fiori apps for measuring points and counters, I recommend referring to the SAP Fiori Apps Library or consulting with your SAP system administrator or SAP representative.

3.38. Customizing Settings for Measuring Points and Counters

This section describes customizing the settings for Measuring Points and Counters (see Table 3-6).

Table 3-6. *IMG Settings for Measuring Points and Counters*

Field Name or Data Type	Menu Path
Measuring Point Category	Plant Maintenance and Customer Service → Master Data in Plant Maintenance and Customer Service → Basic Settings → Measuring Points, Counters and Measurement Documents → Define Measuring Point Categories

3.39. Summary

In this chapter, you learned how SAP Measuring Points and Counters provides a comprehensive understanding of managing measuring points and counters within the SAP solution. It covered creating, configuring, and using measuring points to capture readings from physical assets or equipment. The chapter emphasized the importance of defining measuring point hierarchies and assigning them to relevant objects to monitor key performance indicators (KPIs) and trigger maintenance activities. It explored functionalities such as counter management, enabling tracking usage or production quantities for equipment. The chapter also highlighted integration with other SAP modules, such as maintenance and plant maintenance, to facilitate proactive maintenance planning and optimize asset performance. Overall, this chapter equips users with the knowledge and tools to effectively manage measuring points and counters, monitor equipment performance, and drive maintenance strategies within the SAP environment.

CHAPTER 4

Business Processes in Asset Management

4.1. Standard Maintenance Process

In SAP S/4HANA EAM, the standard maintenance process refers to the established and predefined series of activities that organizations follow to effectively manage their assets and equipment. This process ensures that assets are properly maintained, monitored, and repaired to optimize their performance and lifespan. The standard maintenance process typically encompasses the following key phases.

- **Planning**: This phase involves defining the maintenance requirements for assets, including scheduling maintenance tasks, determining resource needs (e.g., labor, materials, tools), and establishing the necessary budget.

- **Scheduling**: Once the maintenance tasks are planned, they are scheduled based on factors such as asset criticality, availability of resources, and operational impact. This step aims to minimize downtime and disruptions.

© Rajesh Ojha and Chandan Mohan Jaiswal 2023
R. Ojha and C. M. Jaiswal, *SAP S/4HANA Asset Management*,
https://doi.org/10.1007/978-1-4842-9870-1_4

- **Execution**: During this phase, the actual maintenance activities are carried out. It includes inspections, repairs, preventive maintenance, corrective maintenance, and other relevant tasks.

- **Documentation**: Accurate and thorough documentation of maintenance activities is crucial. It includes recording details of work performed, parts used, labor hours, and any issues encountered. Proper documentation helps in historical tracking, compliance, and future decision-making.

- **Completion and Reporting**: The relevant data is entered into the system after the maintenance tasks are completed. It includes updates on asset status, maintenance costs, and any changes to asset condition. These reports contribute to performance analysis and decision-making.

- **Analysis**: Organizations analyze maintenance data to identify trends, patterns, and opportunities for improvement. This step supports decision-making regarding optimizing maintenance strategies, resource allocation, and asset replacement.

- **Optimization**: Organizations refine their maintenance processes and strategies based on the analysis. This might involve adjusting maintenance schedules, improving resource allocation, and implementing best practices to enhance asset performance and reduce downtime.

SAP S/4HANA EAM provides a comprehensive suite of tools and features to streamline and automate these processes. It enables organizations to manage their assets efficiently, reduce operational costs, extend asset lifecycles, and improve overall business performance. The system allows integration with other SAP modules and external systems, providing a holistic view of asset management across the organization.

It's important to note that while the standard maintenance process provides a structured framework, organizations can customize and adapt it to their specific industry, asset types, and business needs.

4.1.1. Modeling Maintenance Process

Various methods are employed to model the maintenance process (see Figure 4-1), contingent on factors such as the nature of the task, the software ecosystem, and the utilization of pre-existing configurations. When responding to unexpected breakdowns or damage, the reactive maintenance process typically operates more swiftly with minimal planning. Conversely, for routine inspections or maintenance, the proactive maintenance process can be pre-planned and executed to a certain degree automatically.

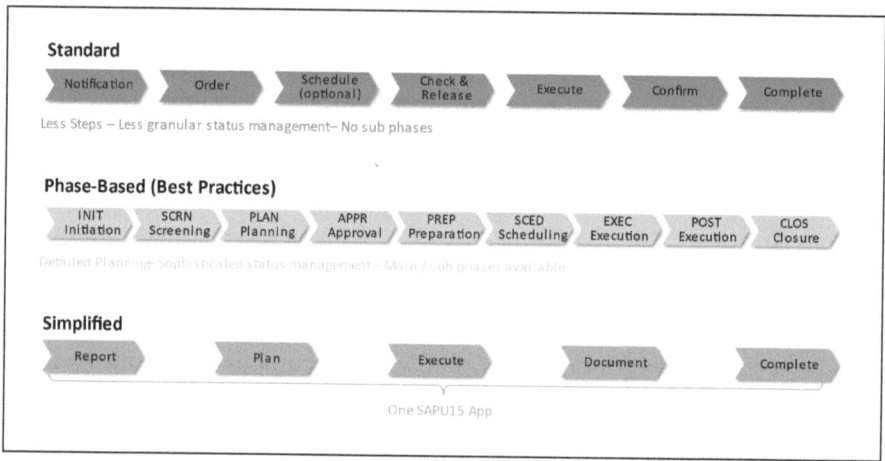

Figure 4-1. *Model maintenance process*

As shown in Figure 4-1, there are three process variations based on diverse needs.

- **Standard process**: This method employs conventional notifications and orders. It offers adaptability and can be adjusted for various demands like sudden failures or planned tasks.

- **Phase-based process**: This approach features meticulous and refined planning. It encompasses multiple phases and subphases, suitable for use in regulated settings. It can be augmented with preconfigured content if associated Best Practices scope items are implemented.

- **Simplified process**: Utilizing a SAPUI5 foundation, this process permits the complete procedure to be executed within a single Fiori app. This version is generally employed for straightforward repair work, where intricate planning isn't necessary.

The standard maintenance process undergoes the following stages.

1. **Create notification.** Malfunctions and other requirements are documented within notifications. Notifications are accessible and manageable through the notification list editing.

2. **Plan maintenance order.** During this stage, orders are generated and scheduled based on reported needs. The planning involves defining necessary steps, required materials, and any pertinent PRTs, such as measurement instruments or cranes.

3. **Check and release maintenance order.** This stage subjects the order to various checks, including material availability and capacity planning, which are crucial for order release. The system examines permit approval, safety plans, and compliance with work clearance criteria in specific areas. If these checks raise no issues, the order progresses into the processing stage. Typically, shop papers are printed at this juncture.

4. **Execute maintenance order.** The actual execution of the order occurs here. The necessary materials, even those not initially planned or reserved, are retrieved for the order's execution.

5. **Confirm maintenance order.** This stage involves inputting time and technical confirmations into the system.

6. **Complete maintenance order.** In this stage, the order is marked as technically complete. The order's settlement by the controlling (CO) department can occur before or after the technical completion. Eventually, the controlling department designates the maintenance order as "business complete."

4.2. Notification

In SAP EAM, notification is a critical element in the maintenance and service management process. This formal communication within the system indicates a situation that requires attention, action, or response related to assets, equipment, or processes. Notifications can initiate various maintenance activities, such as repairs, inspections, preventive maintenance, and more.

The following are some key points about notifications in SAP EAM.

- **Initiation of work**: Notifications are the starting point for maintenance work. When a problem or issue is identified with an asset or equipment, a notification is created to document the issue and trigger the appropriate actions.

- **Information capture**: Notifications contain detailed information about the issue or situation, such as the description of the problem, location of the asset, priority, affected equipment, and more. This information helps maintenance teams understand the nature of the issue and respond accordingly.

- **Types of notifications**: There are different types of notifications in SAP EAM, including maintenance notifications, service notifications, quality notifications, and more. Each type of notification serves a specific purpose and may follow a different workflow.

- **Workflow and processing**: Notifications in SAP EAM typically follow predefined workflows. Once a notification is created, it can be assigned to responsible parties, such as maintenance technicians or teams. The workflow involves assessment, planning, scheduling, execution, and completion of the required maintenance activities.

- **Integration with other processes**: Notifications are closely integrated with other processes in SAP EAM, such as work orders, maintenance plans, and purchase requisitions. When a notification requires maintenance work, it can be converted into a work order to ensure the proper execution of tasks.

- **Tracking and reporting**: Notifications provide a way to track and monitor the progress of maintenance activities. This data can be used for reporting, analysis, and continuous improvement of maintenance processes.

- **Notification types**: Different notification types may have distinct attributes and fields based on their specific purpose. For example, a service notification might include information about a customer issue, while a maintenance notification might focus on an equipment malfunction.

- **Prioritization**: Notifications often include a priority designation to help maintenance teams determine the issue's urgency and prioritize their workload accordingly.

Overall, notifications play a crucial role in SAP EAM by streamlining communication, facilitating efficient maintenance processes, and ensuring that assets and equipment are properly maintained to minimize downtime and optimize performance.

In corporations, the initial step involves generating maintenance needs within the system by creating notifications. This process aims to enhance prioritization and coordination efforts. These notifications must encompass comprehensive data to retain a historical record and facilitate subsequent assessments.

The inception of the corrective maintenance procedure hinges on generating a notification concerning issues such as damage, malfunctions, or modification requests. Generally, this notification pertains to a technical entity and describes the malfunction or requisites. Furthermore, supplementary data can be inputted into the notification to establish a historical record, encompassing factors like damages and causes.

Maintenance tasks can be strategized based on notifications, often leading, though not invariably, to formulating a maintenance order.

4.2.1. Notification Structure

Maintenance notifications (see Figure 4-2) include header data to identify and manage them. This information remains consistent across the entire notification.

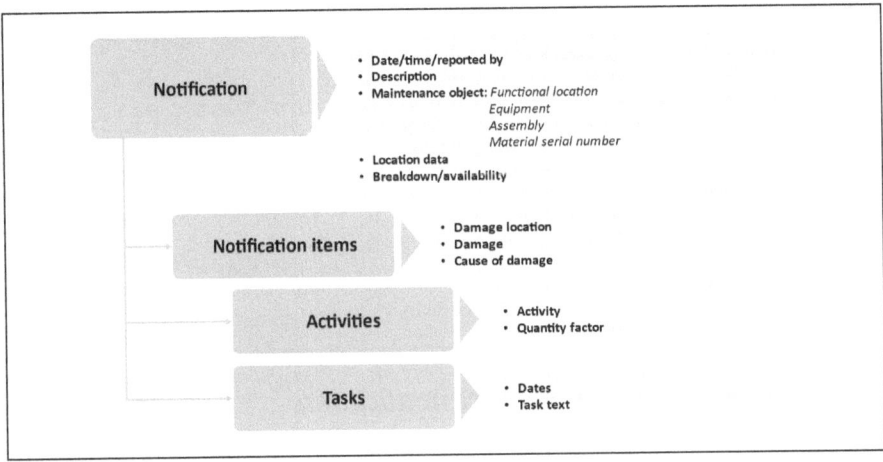

Figure 4-2. *Notification structure*

Data is entered and maintained in notification items to provide a more detailed account of problems, damages, or activities. Notifications (see Figure 4-3) can consist of multiple items.

Activities within a notification document the work carried out, especially crucial for inspections as they are evidence of completed tasks. Tasks, on the other hand, outline pending activities. This encompasses tasks arising after the maintenance task, such as creating a report. In some cases, tasks can also aid in planning, particularly when order processing is inactive. Different individuals can be designated for notification processing during such instances, and task execution can be monitored over specific timeframes. However, it's important to note that certain processes, like cost monitoring, material planning, or capacity requirements planning, are not feasible with this type of processing.

The notification interface is customizable, allowing the adjustment of tabbed pages and their content through Customizing to align with specific requirements.

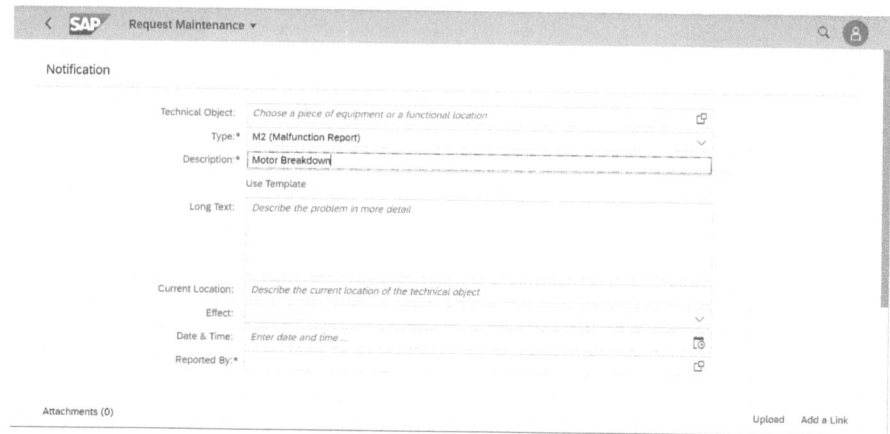

Figure 4-3. *Request maintenance notification*

4.3. Customizing Settings for Notifications

This section describes the Customizing settings for Notifications. Table 4-1 lists and describes IMG Settings for Notifications.

Table 4-1. *IMG Settings for Notifications*

Field Name and Data Type	Menu Path
Overview of Notification Type	Plant Maintenance and Customer Service → Maintenance and Service Processing → Maintenance and Service Notifications → Overview of Notification Type
Action Box	Plant Maintenance and Customer Service → Maintenance and Service Processing → Maintenance and Service Notifications → Notification Processing → Additional Functions → Define Action Box
General Notification	Cross-Application Components → Notification
System Condition	Plant Maintenance and Customer Service → Maintenance and Service Processing → Maintenance and Service Notifications → Notification Processing → Define System Conditions
Catalog Profile	Plant Maintenance and Customer Service → Maintenance and Service Processing → Maintenance and Service Notifications → Notification Creation → Notification Content → Define Catalog Profile
Screen Templates	Plant Maintenance and Customer Service → Maintenance and Service Processing → Maintenance and Service Notifications → Notification Creation → Notification Types → Set Screen Templates for the Notification Type

(*continued*)

Table 4-1. (*continued*)

Field Name and Data Type	Menu Path
Allowed Change of Notification Type	Plant Maintenance and Customer Service → Maintenance and Service Processing → Maintenance and Service Notifications → Notification Creation → Notification Types → Allowed Change of Notification Type
Field Selection	Plant Maintenance and Customer Service → Maintenance and Service Processing → Maintenance and Service Notifications → Notification Creation → Notification Types → Set Field Selection for Notifications
Integration Notification/ Order (Long Text Transfer, Object List Control)	Plant Maintenance and Customer Service → Maintenance and Service Processing → Maintenance and Service Orders → Functions and Settings for Order Types → Define Notification and Order Integration
Catalogs	Plant Maintenance and Customer Service → Maintenance and Service Processing → Maintenance and Service Notifications → Notification Creation → Notification Content → Maintain Catalogs
Long Text Control	Plant Maintenance and Customer Service → Maintenance and Service Processing → Maintenance and Service Notifications → Notification Creation → Notification Types → Define Long Text Control for Notification Types
List Editing	Plant Maintenance and Customer Service → Maintenance and Service Processing → Maintenance and Service Notifications → Notification Processing → List Editing

(*continued*)

Table 4-1. (*continued*)

Field Name and Data Type	Menu Path
Notification Type	Plant Maintenance and Customer Service → Maintenance and Service Processing → Maintenance and Service Notifications → Notification Creation → Notification Types
Assign Notification Types to Order Types	Plant Maintenance and Customer Service → Maintenance and Service Processing → Maintenance and Service Notifications → Notification Creation → Notification Types → Assign Notification Types to Order Types
Number Ranges	Plant Maintenance and Customer Service → Maintenance and Service Processing → Maintenance and Service Notifications → Notification Creation → Notification Types → Define Number Ranges
Object Information	Plant Maintenance and Customer Service → Maintenance and Service Processing → Maintenance and Service Notifications → Notification Processing → Object Information
Partner Determination Procedure	Plant Maintenance and Customer Service → Maintenance and Service Processing → Maintenance and Service Notifications → Notification Creation → Partners
Priorities	Plant Maintenance and Customer Service → Maintenance and Service Processing → Maintenance and Service Notifications → Notification Processing → Response Time Monitoring → Define Priorities

(*continued*)

Table 4-1. (*continued*)

Field Name and Data Type	Menu Path
Response Profile	Maintenance and Customer Service → Maintenance and Service Processing → Maintenance and Service Notifications → Notification Processing → Response Time Monitoring → Define Response Monitoring
Transaction Start Values	Plant Maintenance and Customer Service → Maintenance and Service Processing → Maintenance and Service Notifications → Notification Creation → Notification Types → Define Transaction Start Values
Fiori Apps Settings: Notification Type Planning Plant	Plant Maintenance and Customer Service → Maintenance and Service Processing → Fiori Apps for Maintenance Processing → Assign Notification Types to Maintenance Planning Plants
Fiori Apps Settings: Define Overall Status	Plant Maintenance and Customer Service → Maintenance and Service Processing → Fiori Apps for Maintenance Processing → Define Overall Status

4.3.1. Catalogs

The catalog finds applications while managing notifications, facilitating the systematic input of outcomes and tasks using codes. This coded approach proves especially valuable for analytical purposes (see Figure 4-4). Within the Plant Maintenance Information System (PMIS), specific predefined analyses exist that leverage these codes for in-depth examination.

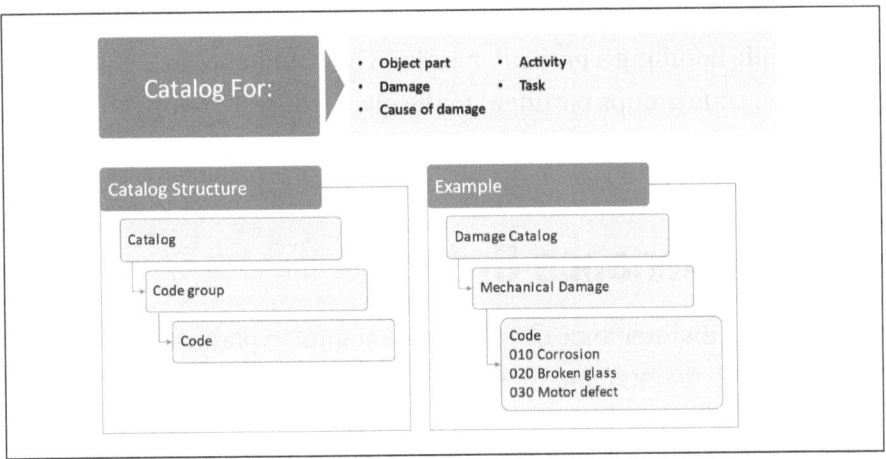

Figure 4-4. *Catalog*

The following describes the catalog's features.

- **Catalog**: This is an aggregation of code groups united by their content (such as types of damages and their causes).

- **Code groups**: These are clusters of codes organized based on their associated content (for instance, damages to vehicles, pumps, and motors or distinctions between mechanical and electrical damage).

- **Codes**: These are descriptions encompassing damages, activities, causes of damage, parts of objects, tasks, etc. This approach substantially minimizes the likelihood of inaccurate entries. Codes can serve as the initial stage for workflows and subsequent processes. Standard analyses within the PMIS enable statistical assessments to be conducted.

Within a catalog profile, it's possible to stipulate the code groups utilized while handling a particular entity. The benefit lies in presenting solely those code groups pertinent to the given entity. Catalog profiles can be linked to either a technical object or a notification type.

4.4. Maintenance Orders

In SAP EAM, a maintenance order is fundamental in planning, executing, and monitoring maintenance activities on assets and equipment. It is a formal document outlining the tasks, resources, materials, and timelines required to perform maintenance work. Maintenance orders play a central role in managing maintenance processes efficiently and effectively.

The following are the key aspects of maintenance orders in SAP EAM.

- **Planning and scheduling**: Maintenance orders are created based on notifications, work requests, or maintenance plans. They provide a structured way to plan and schedule maintenance tasks, ensuring that the right resources are allocated and available when needed.

- **Work execution**: A maintenance order contains detailed information about the work to be performed, including the scope of tasks, labor requirements, required tools and materials, safety instructions, and other relevant information. This information guides technicians during the execution of maintenance activities.

- **Resource allocation**: Maintenance orders allow allocating personnel, equipment, and materials to specific tasks. This ensures the necessary resources are organized and utilized efficiently, reducing downtime and improving productivity.

- **Cost tracking**: Maintenance orders facilitate cost tracking by capturing labor hours, materials used, and other expenses incurred during the maintenance process. This data helps organizations analyze maintenance costs and make informed decisions about resource allocation.

- **Status tracking**: Throughout the maintenance process, the order's status is updated to reflect its progress, from planning and scheduling to execution and completion. This real-time status tracking enables better communication and coordination among teams.

- **Integration with notifications**: Maintenance orders often originate from notifications. Once a maintenance notification is created and assessed, it can be converted into a maintenance order to initiate the work.

- **Measurement and documentation**: Maintenance orders allow for recording measurements, observations, and notes related to the maintenance activities. This documentation helps maintain a historical record of work performed and can be useful for future reference or audits.

- **Follow-up actions**: After maintenance work is completed, the order can trigger follow-up actions, such as creating service reports, updating asset history, or generating maintenance history for reporting purposes.

- **Integration with financial processes**: Maintenance order data, including costs incurred, can be integrated with financial systems for accurate cost allocation, budgeting, and reporting.

- **Performance evaluation**: Analyzing maintenance orders and their associated data enables organizations to evaluate their maintenance strategies' effectiveness, identify improvement areas, and optimize maintenance processes.

In summary, maintenance orders in SAP EAM provide a structured framework for planning, executing, and managing maintenance tasks. They help organizations streamline their maintenance processes, reduce downtime, manage costs, and maintain assets and equipment to achieve optimal operational performance.

4.4.1. Corrective/Reactive Maintenance Process

The effective execution of maintenance tasks on a technical object necessitates meticulous pre-planning to enhance efficiency and optimize resource utilization. Subsequently, a maintenance order (see Figure 4-5) is orchestrated from a notification, constituting the second phase of the process. Typically, planning activities encompass formulating order operations, predetermined work efforts, and preserving pertinent materials or spare parts.

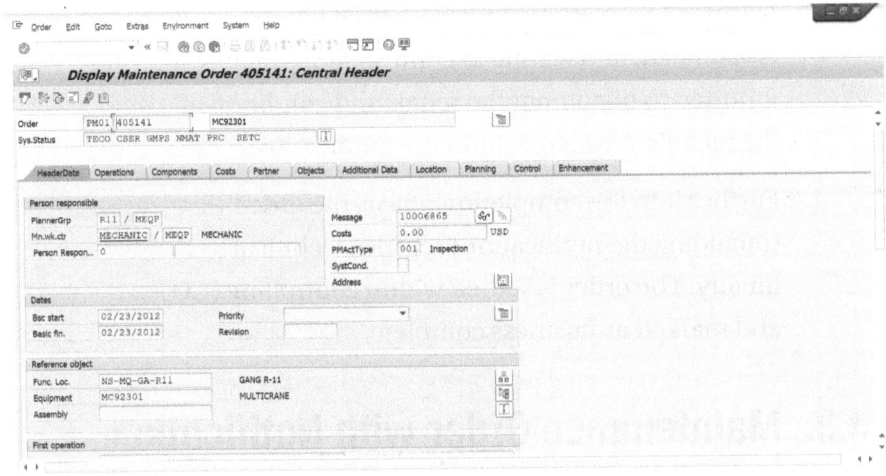

Figure 4-5. *Display maintenance order*

4.4.2. Breakdown Maintenance Process

Breakdown maintenance involves the instantaneous generation and authorization of a maintenance order, condensing the process into a single step right after a malfunction has been reported, often by a production employee. Depending on the procedural model, this step can be done through the SAP GUI (transaction) or via the web (SAP Fiori launchpad or Business Client). Alternatively, the notification can be initiated using a mobile device.

The breakdown maintenance process consists of the following steps.

1. The process initiates with a maintenance order's inception and a malfunction notification. Subsequently, the maintenance order is set into motion.

2. The execution phase involves retrieving spare parts from inventory and the direct execution of the order.

3. During the completion phase, actual time expended is validated, and technical confirmations are inputted to document the repair and condition of the technical system within the notification.

4. Further into the completion phase, the order (including the notification) reaches technical finality. The order is settled within controlling (CO) and marked as business complete.

4.4.3. Maintenance Order with Notification

The following describes the tools accessible for managing maintenance tasks.

- Orders serve to strategize maintenance tasks and plan and oversee incurred costs. It's not obligatory to execute planning functions; orders can also be generated as immediate orders devoid of planning.

- Notifications are employed to communicate maintenance requisites, document technical findings, and record performed activities. Orders and notifications can be utilized distinctly. Nevertheless, they are frequently integrated to leverage the benefits of both tools, as shown in Figure 4-6.

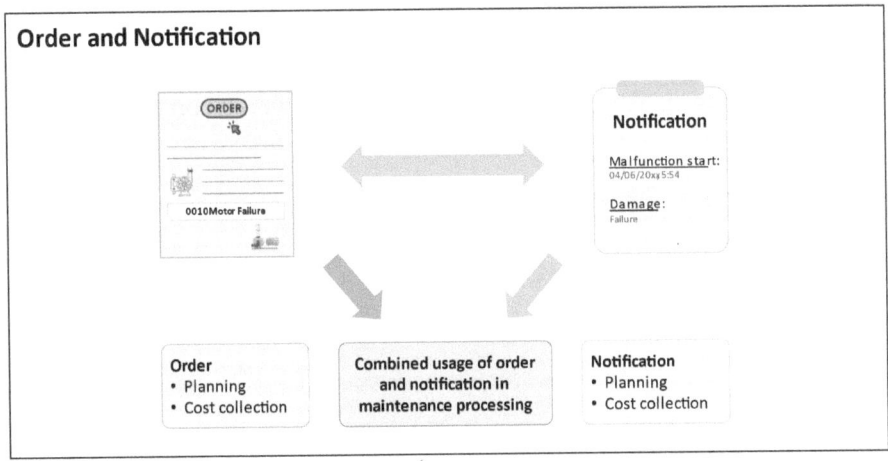

Figure 4-6. *Order and notification*

In approach A, the initial stage involves crafting a malfunction notification (see Figure 4-7) containing pertinent particulars to delineate the malfunction. Subsequently, a breakdown order is formulated in connection with the malfunction notification. Typically, the order pertains to a technical entity (e.g., a functional location or equipment). In the initial order operation, the malfunction is briefly described.

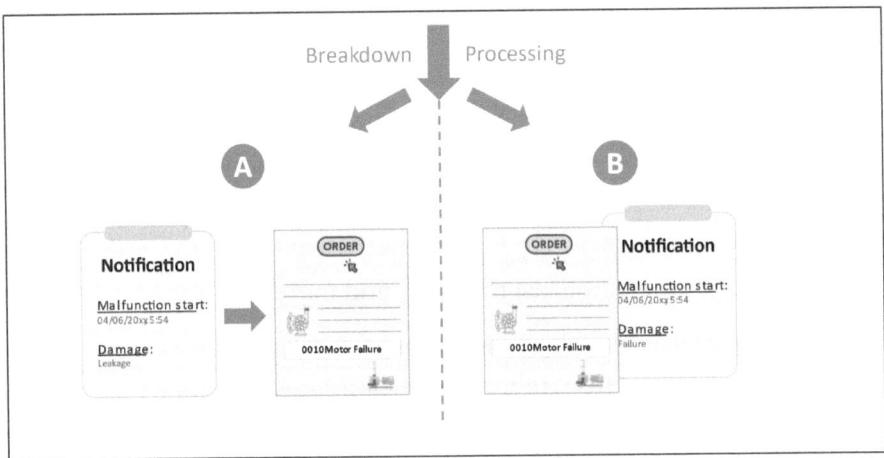

Figure 4-7. *Approaches to process maintenance order*

Alternatively, approach B shows a breakdown order can be directly initiated alongside a malfunction notification. This method's advantage lies in the simultaneous display of both the Report and Order tabs on the interface.

4.4.4. Order Structure

Figure 4-8 illustrates the core components of a maintenance order.

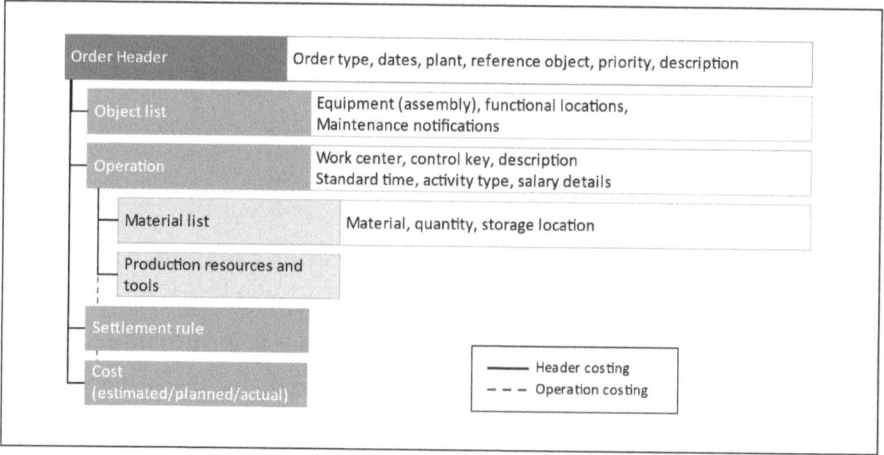

Figure 4-8. *Maintenance order structure*

- **Order header**: The order header data includes information for identifying and managing the maintenance order. This data holds throughout the entire order and comprises details like order number, description, order type, scheduled execution dates, task priorities, creator, last modifier, and more.

- **Object list**: The object list enumerates the entities to be processed (e.g., functional locations, equipment, assemblies, serial numbers). It is utilized when the same activity needs to be executed across multiple objects of the same kind.

- **Operation**: Order operations outline the tasks to be accomplished, the individuals responsible for these tasks, and the guidelines they must adhere to.

- **Materials list (components list)**: This list records spare parts utilized during order execution.

- **Production resources and tools**: Resources such as tools, protective gear, and vehicles that are essential for order execution, are necessary but not consumed as they can be reused.

- **Settlement rule**: The settlement rule data specifies the account assignment object for settling costs. The suggested account assignment object comes from the reference object's master record, which can be changed when the initial settlement rule is established for the order.

- **Cost**: The cost view presents the order's estimated, planned, and actual costs across value categories. Both a technical view and a controlling view are accessible.

The following are options for creating maintenance orders.

- Scenario 1: Orders can be created with a single operation (quick entry) or more complex planning with multiple operations. Operations can be listed in the order's operations list.

- Scenario 2: Direct creation of a maintenance order (e.g., a breakdown order).

- Scenario 3: Maintenance notification is submitted by a requester, and a maintenance order is then created by the planner referencing the notification.

- Scenario 4: A maintenance order can feature multiple technical objects in the Object List tab, with individual notifications assigned to each object.

- Scenario 5: A maintenance order can be generated without referencing a maintenance notification. A technical confirmation in the form of an activity report is created for this order.

- Scenario 6: A maintenance order can be automatically generated from a maintenance item within a maintenance plan.

4.5. Customizing Settings for Maintenance Order

This section describes the Customizing settings for Maintenance Order. Table 4-2 lists and describes IMG Settings for Maintenance Order.

Table 4-2. *IMG Settings for Maintenance Order*

Field Name and Data Type	Menu Path
Configure Order Types	Plant Maintenance and Customer Service → Maintenance and Service Processing → Maintenance and Service Orders → Print Control → Functions and Settings for Order Types → Configure Order Types
Number Ranges	Plant Maintenance and Customer Service → Maintenance and Service Processing → Maintenance and Service Orders → Functions and Settings for Order Types → Configure Number Ranges

(continued)

Table 4-2. (*continued*)

Field Name and Data Type	Menu Path
Planning Plant	Plant Maintenance and Customer Service → Maintenance and Service Processing → Maintenance and Service Orders → Functions and Settings for Order Types → Assign Order Types to Maintenance Plants
Costing Variant	Plant Maintenance and Customer Service → Maintenance and Service Processing → Maintenance and Service Orders → Functions and Settings for Order Types → Costing Data for Maintenance and Service Orders → Assign Costing Parameters and Results Analysis Keys
Settlement Rule	Plant Maintenance and Customer Service → Maintenance and Service Processing → Maintenance and Service Orders → Functions and Settings for Order Types → Costing Data for Maintenance and Service Orders → Settlement Rule: Define Time and Creation of Distribution Rule
Maintenance Activity Type	Plant Maintenance and Customer Service → Maintenance and Service Processing → Maintenance and Service Orders → Functions and Settings for Order Types → Costing Data for Maintenance and Service Orders → Maintenance Activity Type
Planning Indicator	Plant Maintenance and Customer Service → Maintenance and Service Processing → Maintenance and Service Orders → Functions and Settings for Order Types → Costing Data for Maintenance and Service Orders → Define Default Value for Planning Indicator for Each Order Type

(*continued*)

Table 4-2. (*continued*)

Field Name and Data Type	Menu Path
Profiles	Plant Maintenance and Customer Service → Maintenance and Service Processing → Maintenance and Service Orders → Functions and Settings for Order Types → Costing Data for Maintenance and Service Orders → Create Default Value Profiles for External Procurement .../Create Default Value Profiles for General Order Data
Change Docs/Purchase Requisition/MRP	Plant Maintenance and Customer Service → Maintenance and Service Processing → Maintenance and Service Orders → Functions and Settings for Order Types → Costing Data for Maintenance and Service Orders → Define Change Documents, Collective Purchase Requisition, MRP relevance
Priorities	Plant Maintenance and Customer Service → Maintenance and Service Processing → Maintenance and Service Orders → General Data → Define Priorities
Partners	Plant Maintenance and Customer Service → Maintenance and Service Processing → Maintenance and Service Orders → Partner
Documentations of Good Movements	Plant Maintenance and Customer Service → Maintenance and Service Processing → Maintenance and Service Orders → Goods Movements for Order
Object Info	Plant Maintenance and Customer Service → Maintenance and Service Processing → Maintenance and Service Orders → Object Information

4.5.1. Controlling

Let's compare cost calculations and postings between the header level and the operation level.

By default, the computation of costs for maintenance orders occurs at the header level. However, the operation account assignment (OAA) solution introduces the capability to compute maintenance order costs at the operation level. The summation of header totals is performed dynamically as needed, with no cost storage on the OAA order object database.

Similarly, when it comes to cost postings, conventional practice involves settling costs for maintenance orders at the header level. However, the OAA approach empowers you to post maintenance order costs at the operation level. A maintenance order must adopt either header-based or operation-based costing; a mixed-mode costing approach is not permissible. While certain maintenance order types, such as PM01 and PM02, predominantly employ header-level postings, there arises a need to both compute and post costs at the operation level.

To activate OAA or operation-level costing (OLC), you configure the Customizing settings for a specific combination of maintenance order type and the maintenance planning plant.

It's crucial to note that an order must unequivocally adopt either header-based or operation-based costing, with mixed-mode costing being infeasible.

Ordinarily, orders subjected to header-based costing automatically generate their settlement rule based on the account data of the reference object present in the order header.

Conversely, orders employing OAA adhere to a similar logic—relying on the reference object within the order header—but each operation creates an individual settlement rule.

In scenarios where a technical object is linked to an operation, the settlement rule for said operation takes its cues from the data associated with this object.

4.5.2. Planning Material and Services

When creating a maintenance order in SAP EAM, you can plan and allocate both internal and external materials and services to ensure the successful execution of maintenance tasks. Let's look at how to do it.

The following steps address *internal materials.*

1. In the maintenance order creation screen, navigate to the Materials tab or a similar section dedicated to materials.

2. Select the option to add internal materials.

3. Search for and select the required materials from your organization's material master or inventory.

4. Specify the quantities and units needed for the maintenance task.

5. Assign the materials to the appropriate order operations if the order contains multiple operations.

6. The system may provide information about the availability of materials based on your organization's inventory.

The following steps address *external services.*

1. On the maintenance order creation screen, navigate to the Services tab or an equivalent section.

2. Add external services by searching for the required services using a service catalog or predefined service entries.

3. Provide details such as the service description, quantity, unit of measure, and specific requirements.

4. Assign the services to the relevant order operations if necessary.

The following steps address *external materials.*

1. External materials are typically handled similarly to internal materials but might require additional procurement steps.

2. You can create a purchase requisition or purchase order to procure the required external materials from vendors.

The following steps address *service entry sheets.*

1. After the maintenance work is completed and the external services are provided, you can create a service entry sheet to verify the services rendered by external vendors.

2. The service entry sheet confirms the completion of the services and initiates the invoice verification process.

The following steps address *integration with procurement.*

1. SAP EAM can be integrated with procurement modules like SAP MM (Materials Management) and SAP SRM (Supplier Relationship Management).

2. For external materials and services, the procurement process can involve creating purchase requisitions, purchase orders, and tracking the procurement lifecycle.

The following steps address *cost tracking and settlement.*

1. By planning and allocating materials and services to maintenance orders, you facilitate accurate cost tracking.

2. After execution, the actual costs incurred for materials and services can be compared to the planned costs.

3. Settlement processes can be used to allocate costs from maintenance orders to cost centers or other relevant accounts.

Finally, let's address *scheduling and availability checks.* During the planning phase, the system might perform an availability check as shown in Figure 4-9 to ensure that the required materials are available in stock or can be procured in time. This helps prevent delays due to material shortages.

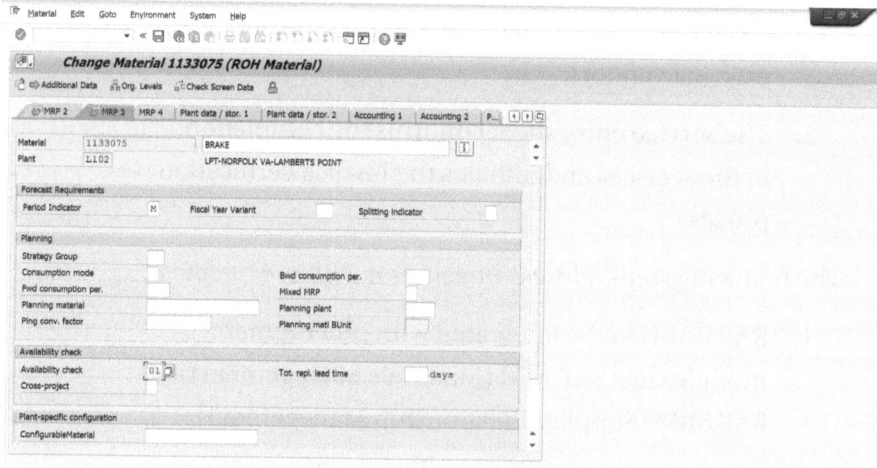

Figure 4-9. Material availability check

It's important to note that the specific steps and terminology could vary slightly based on your organization's SAP EAM configuration and version. Always refer to the documentation and training materials specific to your SAP system for accurate guidance.

4.6. Customizing Settings for Material Planning

This section describes the Customizing settings for Material Planning. Table 4-3 lists and describes IMG Settings for Maintenance Planning.

Table 4-3. *IMG Settings for Maintenance Planning*

Field Name and Data Type	Menu Path
Material Availability Check	Plant Maintenance and Customer Service → Maintenance and Service Processing → Maintenance and Service Orders → Functions and Settings for Order Types → Availability Check for Material, PRTs and Capacities
Interface for Procurement Using Catalogs (OCI)	Plant Maintenance and Customer Service → Maintenance and Service Processing → Maintenance and Service Orders → Interface for Procurement Using Catalogs

4.6.1. Resource Scheduling for Maintenance Planners

SAP S/4HANA offers an additional feature for scheduling resources, known as SAP S/4HANA Asset Management for Resource Scheduling (RSH), as shown in Figure 4-10.

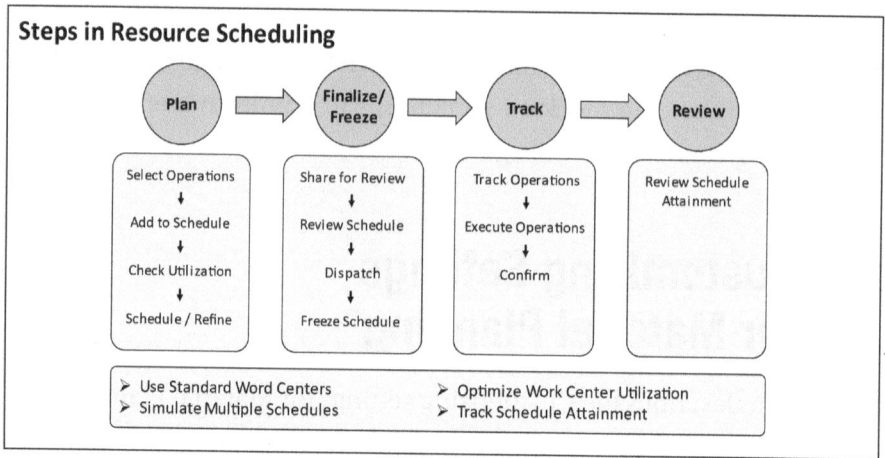

Figure 4-10. *Resource scheduling steps*

The Fiori group titled Resource Scheduling for Maintenance Planner includes the following tiles.

- Resource Scheduling for Maintenance Planners

- Schedule Management

- Maintenance Scheduling Board

- Allocation of Maintenance Order Operations

- Schedule Management - Under Review Fiori App

The Fiori Resource Scheduling for Maintenance Planners application provides several functions or cards.

- Outstanding maintenance orders prioritized by the due date, featuring operations scheduled within the upcoming four weeks

- Outstanding maintenance orders with operations scheduled within the upcoming four weeks

- Work center usage based on maintenance plans and order operations

- Pending maintenance orders with operations concluded over the previous six months

- Unassigned work

- Schedules

4.6.2. Utilization Analysis for Work Centers

Within the Utilization Analysis app for work centers, you have the capability to assess the capacity status of your work centers. The utilization chart empowers you to display utilization based on several attributes, including the following.

- Work center

- Priority

- Order type

- Activity type

- Processing status

- Control key

Furthermore, you can perform the following actions within the app.

- Schedule, modify, and dispatch order operations as well as suboperations

- Possibility to cancel dispatch operations

- To balance work center utilization, you can alter the work center assigned to order operations and suboperations

- Adjust the start date and time for order operations

The app provides the following functionalities.

- Setting specific start dates for order operations

- Modification of the processing status of order operations, with visualization through distinct colors

- Simulation feature for rescheduling planned dates

- Dispatching of scheduled order operations, incorporating simulated outcomes

- Availability of multiple views for utilization forecasting, organized according to work centers and days

4.7. Customizing Settings for Order Scheduling

This section describes the Customizing settings for Order Scheduling. Table 4-4 lists and describes IMG Settings for Order Scheduling.

Table 4-4. *IMG Settings for Order Scheduling*

Field Name and Data Type	Menu Path
Scheduling	Plant Maintenance and Customer Service → Maintenance and Service Processing → Maintenance and Service Orders → Scheduling
Time Zone Support	Plant Maintenance and Customer Service → Maintenance and Service Processing → Basic Settings → Activate Time Zone Support for Application Areas

4.7.1. Checking and Release Maintenance Order

The maintenance planner holds the responsibility of ensuring the timely processing of orders. This entails guaranteeing the availability of materials, generating shop papers, and authorizing orders for further processing.

This phase within the corrective maintenance process encompasses the following actions.

- Selection of maintenance orders earmarked for execution

- Examination of material availability

- Capacity assessment for available resources

- Assessment and issuance of relevant permits

- Risk evaluation and formulation of a safety plan

- Tagging of pertinent areas (applies when using WCM)

- Release of orders and printing of shop papers

The following are some key aspects.

- Reviewing outstanding notifications yet to be processed and unassigned to maintenance orders.

- Analyzing maintenance orders that remain in the planning phase without being released for processing.

- Assessing unreleased purchase requisitions or purchase orders for non-stock materials essential as spare parts in maintenance orders.

- Displaying approved purchase requisitions for non-stock materials lacking corresponding purchase orders.

- Identifying non-stock materials ordered but potentially unavailable by the required date.

- Evaluating released maintenance orders that have passed their end date but are still awaiting final confirmation.

- Analyzing confirmed maintenance orders whose stipulated end date falls within the chosen reference period but remain incomplete both technically and from a business standpoint.

The following methods are used for material availability verification.

- Verification via order list for multiple orders.

- Verification through a background job for a considerable number of orders.

- Individual order verification.

For stock materials planned to be used in order operations, the system can conduct a comprehensive availability check in one step. This function, known as the Availability Check, verifies whether all materials in the maintenance order are sufficiently available. Depending on system settings and data entered in material master records, the system carries out an availability check for all materials linked to operations within the maintenance order. The system promptly notifies you of the outcome through an online message. In the event of scarcity, the system presents an error log containing detailed check results.

Upon releasing a maintenance order, the system performs an availability check for planned materials per your customizing configurations. If certain planned materials are insufficiently available, you may release the order if your system settings permit.

Worker safety within EAM involves the following aspects.

- Ensuring worker safety (Web functions)

- Integration with the risk management system (GRC)

These functionalities enable the establishment of a secure work environment by stipulating safety protocols (such as protective gear and safety briefings). These protocols can be linked to maintenance order operations. The cumulative effect of all safety protocols associated with a maintenance order constitutes the safety scheme.

4.7.2. Maintenance Order Release

Upon releasing a maintenance order, the system verifies the accessibility of materials, production resources, and tools. Material reservations become pertinent to material planning no later than the release point, facilitating the withdrawal of materials and the initiation of purchase requisitions.

After order release, the following tasks can exclusively be executed.

- Printing shop papers

- Material withdrawal

- Recording goods receipts (GRs)

- Inputting time confirmations

- Finalizing the task

4.7.3. Maintenance Order Printing

The act of generating shop papers yields the following outcomes.

- **Job ticket**: This document provides a comprehensive overview of the maintenance order for the personnel engaged in the maintenance task. If your system is linked to the document management system (DMS), graphic elements can also be included on the job ticket, such as engineering or design drawings of the pertinent technical system.

- **Operation control ticket**: The operation control ticket furnishes the designated maintenance engineer with a comprehensive outline of the maintenance order, incorporating permit information.

- **Material pick list**: The material pick list guides the warehouse clerk in identifying the materials designated for each operation within the order.

- **Object list**: The object list presents an inclusive view of the technical objects and notifications involved in the order.

- **Time ticket**: For operations adhering to the specific control key, the time ticket encompasses the standard time and duration for the order operations. It is produced only when manual workers are involved, and each worker's time required for executing the operation is recorded on the ticket.

- **Completion confirmation slip**: Utilized by workers, this slip serves as a record of their work times.

- **Material issue slip**: Issued to the maintenance workers, this slip authorizes the retrieval of necessary materials from the warehouse. A separate material issue slip is printed for each material component.

- **Delta printing**: Through delta printing, all unprinted shop papers for a maintenance order can be generated collectively. This functionality is applicable only if the required Customizing setting is configured by your system administration. The outcomes of delta printing encompass.

- Display of new operations (those not previously printed) on the job ticket and operation control ticket.

- Printing of time tickets only if they haven't yet been marked as Printed.

- Printing of components solely if they haven't been previously printed on a component slip, such as the material withdrawal slip.

- Identification of printouts as delta printouts. Once shop papers are printed for a maintenance order, the system automatically designates the order status as Printed and generates a print log.

- **Print log usage**: The print log serves to determine the following.

  - Shop papers already printed for a given maintenance order

  - Initiator of the printing process

  - Timing of the print activities

- **Internal Transaction (IW3D)**: An internal transaction (IW3D) exists for employees who possess the printing capability for orders but lack authorization to modify the overall order.

4.8. Customizing Settings for Maintenance Order Printing

This section describes the Customizing settings for Maintenance Order Printing. Table 4-5 lists and describes IMG Settings for Maintenance Order Printing.

163

Table 4-5. IMG Settings for Maintenance Order Printing

Field Name and Data Type	Menu Path
Define Shop Papers	Plant Maintenance and Customer Service → Maintenance and Service Processing → Maintenance and Service Orders → Print Control → Define Shop Papers, Forms and Output Programs
Define Printer/Print Diversion	Plant Maintenance and Customer Service → Maintenance and Service Processing → Maintenance and Service Notifications → Notification Processing → Additional Functions → Define Action Box
Print Control for Job Card	Plant Maintenance and Customer Service → Maintenance Roles → Maintenance Worker → Configure Print Control for Job Card

4.8.1. Execute Maintenance Orders

During the corrective maintenance process's implementation stage, spare parts are retrieved from the warehouse, and the actual order execution takes place. Manual workers access materials from the warehouse to conduct maintenance activities.

Two withdrawal approaches exist.

- Intended retrieval of stock materials

- Unscheduled retrieval of stock materials

Alternatively, materials can be obtained externally. The movement of goods for a maintenance order is traceable within the order's document flow. To assess the planned and unplanned withdrawals for a material, the material's usage list (IW13) is available for reference.

4.8.2. Confirming Notifications and Orders

Before a maintenance task can be considered technically concluded, the input of working times is registered in the time confirmation, and details encompassing activities, damages, and causes of damage are logged in the technical confirmation. Figure 4-11 shows the confirmation screen.

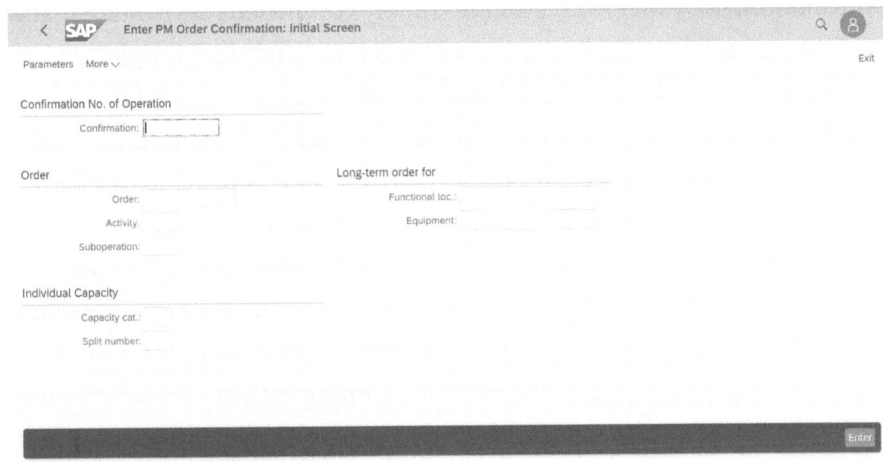

Figure 4-11. *Enter confirmation*

4.8.3. Time Confirmation

There are various methods to validate the time needed for work on a maintenance order.

- Collective input through direct entry or operation list utilization

- Comprehensive completion confirmation for times, activities, measurement values, etc., via a combined interface

- Entry facilitated by the cross-application time sheet (CATS)

Upon inputting completion confirmations for operations or suboperations within a maintenance order, the system automatically designates a partially confirmed (PCNF) status for these operations or suboperations. The Configuration settings under Customizing can trigger an automatic proposal for final completion confirmation instead. Once all operations or suboperations within a maintenance order receive full confirmation, the order itself attains a finally confirmed (CNF) status. The possibility of errant assignment or incorrect data entry for completion confirmations exists. Therefore, the system permits the reversal of completion confirmations when necessary. If multiple orders, each with multiple operations, are confirmed with errors, a mass reversal can be executed from the confirmation list (transaction IW47).

Activities conducted as maintenance notifications can be recorded using the activity report notification type and linked to the corresponding order. Alternatively, confirmation texts can be entered along with time confirmations. However, confirmation texts might lack the same structural organization and analytical ease as activity reports. Measurement values and counter readings are input as measurement documents for the reference object. Technicians employ the confirmation list to validate their orders, often employing a preset query (referred to as a selection variant) that typically employs the operation's work center as a selection criterion. The confirmation list furnishes two confirmation methods.

- Confirm as Planned is direct confirmation without additional processing steps (Confirm function).

- Confirm by Entering/Adjusting Actual Times allows editing and confirmation of actual times (Edit and Confirm function).

4.9. Customizing Settings for Completion Confirmations

This section describes the Customizing settings for completion confirmations. Table 4-6 lists and describes IMG Settings for Completions Confirmations.

Table 4-6. *IMG Settings for Completions Confirmations*

Field Name and Data Type	Menu Path
Control Parameters for Completion Confirmation	Plant Maintenance and Customer Service → Maintenance and Service Processing → Maintenance and Service Orders → Completion Confirmations → Define Control Parameters for Completion Confirmations
Screen Templates for Completion Confirmation	Plant Maintenance and Customer Service → Maintenance and Service Processing → Maintenance and Service Orders → Completion Confirmations → Set Screen Templates for Completion Confirmation

4.9.1. Technical Confirmation

When maintaining technical objects, the most comprehensive technical findings serve as the foundation for subsequent evaluations. Technical findings can encompass the following details.

- Causes of damage

- Performed work (activities and tasks)

- Specific damages and their locations

- Equipment downtimes and system availability during and post-maintenance (system availability)

Technical findings can be recorded either within the malfunction report (which constitutes the basis of the order, if present) or within an activity report created after the order's establishment. Upon completing the maintenance notification, data is transferred to the notification history. This history forms a component of the maintenance record, housing information pertinent to each technical object (e.g., damage, malfunctions, causes, findings, and performed maintenance work).

4.9.2. Complete Notifications and Orders

Once a maintenance task is confirmed, both the order and the notification are subsequently marked as completed. The process of order completion consists of two distinct stages: technical completion and business completion.

- **Technical completion**: A maintenance order is considered technically complete when there are no further tasks left to be fulfilled from a maintenance standpoint.

- **Settlement and business completion**: The Controlling function is responsible for settling maintenance orders and designating them as business complete. This signifies the final step within the corrective maintenance business process. The range of permissible business processes available is notably limited.

- **Technical completion and order**: For the technical completion of a maintenance order, the following options are available.

- Completing the maintenance order and notification individually

- Simultaneously completing the maintenance order along with the associated notifications

Upon completing the maintenance order, its status is altered to TECO (technically completed). Essentially, this denotes fulfilling all maintenance-related tasks outlined by the order. Once a maintenance order achieves the TECO status, further modifications can only be made in specific ways, including the following.

- Locking or unlocking the order

- Applying a deletion flag

- Entering confirmations, invoicing receipts, and outstanding goods movements

- Modifying the settlement rule

If no settlement rule has been established for the maintenance order, the system automatically generates one. In cases where data is insufficient for this process, the system guides you to the point where you can create the settlement rule. Any purchase requisitions lacking corresponding purchase orders are flagged for deletion. Open reservations and capacities are likewise closed.

Business Completion

In SAP maintenance orders, business completion signifies the comprehensive fulfillment of all tasks, resources, and objectives outlined within the order, validating successful execution, and enabling accurate performance evaluation and historical documentation. This milestone ensures operational efficiency and informed decision-making for future maintenance endeavors.

Technical Completion Data and Notification

During technical completion, a reference date and time must be provided, contingent on the periods assigned to the order in the PMIS. This reference date does not impact the determination of location and account assignment data; these details are established based on the order's creation date.

If, for instance, there are changes to the equipment's cost center during order processing, the Update Reference Object Data entry in the context menu can be utilized to effect these changes.

The maintenance history incorporates order data, data from maintenance notifications, and usage histories. This collective data pool serves for retrospective assessment of past work and the formulation of future.

To conclude both the order and notification together, there must be no pending tasks within the notification. If outstanding tasks (marked as OSTS) are present in a notification, completion is hindered until these tasks are resolved. The order linked to the notification can be completed even in the presence of outstanding tasks, as these tasks might not necessarily pertain to the performed order (a new order could be needed in certain instances). The notification status is updated to NOCO (notification completed) upon completion.

Reversing the TECO status is feasible, should the need arise. Reverting a technical completion restores the order to its status before technical completion, recalibrates capacity requirements and reservations, and resets the deletion indicator for unconverted purchase requisitions.

Completing a Notification

Before finalizing a maintenance notification, several factors should be considered.

- Availability and accuracy of data pertaining to the notification's reference object

- Availability and accuracy of relevant item data

- Availability and accuracy of pertinent task data

- Completion or release of all tasks without any outstanding items

- Availability and accuracy of technical data related to the technical object's breakdown and availability

Upon the completion of a maintenance notification, the following consequences ensue.

- The reference date and time define the periods assigned to the notification in PMIS.

- The maintenance notification becomes locked for alterations, rendering further changes to notification data impossible.

- The notification status is updated to NOCO.

4.10. Simplified Maintenance Process

Malfunction Reporting and Repair

This application assists maintenance technicians throughout the breakdown, guiding them from the initial stages to the conclusion in a responsive design, as shown in Figure 4-12.

171

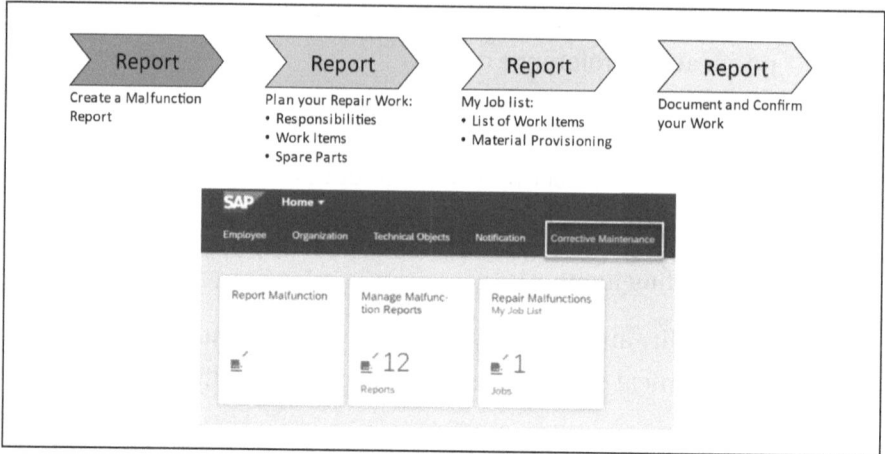

Figure 4-12. *Report and repair malfunction*

The following are the supported steps in the process.

1. Generate a malfunction report encompassing all necessary and pertinent information.

2. Locate pre-existing malfunction records within a comprehensive list.

3. Strategize the repair process, considering all required resources.

    a. Work items designated to the responsible work center

    b. Spare parts needed for the repair task

4. Document and confirm the repair work.

5. Finalize the malfunction report.

The following explains how to create a malfunction report.

1. Identify the affected technical object.

2. Access technical object details by navigating to the master data or the Asset Viewer.

3. Review a list of recently generated malfunction reports concerning the relevant technical object.

4. Draft a detailed description of the issue using an extended text.

5. Specify the present location of the technical object.

6. Choose an effect from the drop-down menu to evaluate the significance of the repair work, such as its impact on safety or environmental compliance.

7. Utilize the camera icon to capture an image of the damage, which is automatically linked to the notification (applicable only on mobile devices).

8. Include a URL for supplementary information.

The following explains how to plan the repair work.

1. Allocate responsibilities to tasks.

2. Define new work items or modify existing ones.

3. Attach spare parts to the designated work items.

4. Add work items.

    a. Input the corresponding work center.

    b. Specify the responsible person for task completion.

    c. Describe the nature of the work to be conducted.

    d. Estimate the required time.

5. Assign spare parts for the repair work.

   a. Search within the technical object's bill of materials.

   b. Browse a list of recently used parts commonly required for repairs on this technical object.

   c. Utilize a comprehensive search encompassing all materials.

   d. Evaluate whether the material is available at a suitable storage location.

The following steps address repairing malfunctions in the My Jobs List.

1. Display a comprehensive list of all tasks assigned to you or your team.

2. Configure the display of jobs within the work list.

   - Mine only: Tasks linked to your number

   - My team's only: Tasks for other technicians within your work center

   - Mine and my team's: All tasks for your work center technicians

   - To be assigned in my team: Unassigned tasks within your work center

3. Apply filters to narrow down the scope of the list based on factors like status or priority.

   In Material Provisioning

   - Inventory Manager executes the goods issuance of the reserved material.

   - Maintenance Technician displays the barcode.

The following explains how to document and confirm work.

1. Validate the malfunction duration.

   a. Confirm the start of the malfunction.

   b. Determine the end of the malfunction.

   c. Indicate whether the machine experienced a breakdown.

2. Record malfunction details.

   a. Specify the impacted object parts.

   b. Select relevant damage codes.

   c. Include cause codes accompanied by explanations.

   d. Enter executed activities.

3. Confirm job details.

   a. Initiate the job.

   b. Pause the job with the reasons provided.

   c. Captured job time is proposed for use.

   d. Confirm the work item.

The Comprehensive Status - An Innovative Status Concept

- Reveals the specific phase of individual tasks

- Chronicles the advancement of the complete malfunction report

- Merges system status and user status into a unified overall status

4.11. Phase-Based Maintenance Process

A holistic phase-based process facilitates technical object maintenance. This comprehensive phase-based procedure provides invaluable support in managing technical objects' maintenance. The process encompasses the treatment of maintenance requests and orders across nine distinct phases, as shown in Figure 4-13.

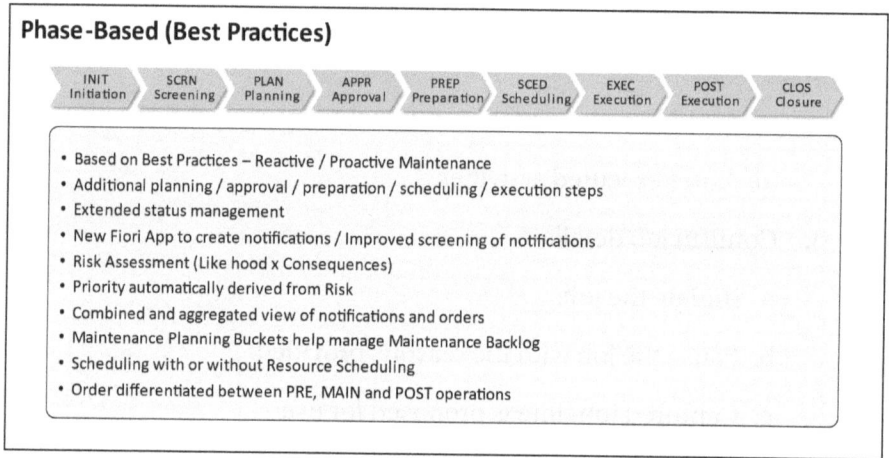

Figure 4-13. *Phase-based maintenance*

For both reactive maintenance and proactive maintenance order types, additional planning, approval, preparation, scheduling, and execution steps are executed, all reflected in supplementary system statuses.

The maintenance process accommodates the following order types.

- **Reactive maintenance** empowers you to conduct maintenance on technical objects when a breakdown or malfunction occurs. This approach minimizes asset downtime, enhancing productivity.

- **Proactive maintenance** enables preemptive measures to avert asset failures or breakdowns. Implementing preventive maintenance and proactive strategies ensures optimal asset utilization and availability. This results in improved asset performance and cost reduction by curtailing breakdowns.

4.11.1. Phases

For both the reactive maintenance and proactive maintenance order types, an enriched series of planning, approval, preparation, scheduling, and execution phases occur, manifesting in supplemental system statuses.

The first phases are initiation and screening.

- **Initiation phase**: Maintenance requests are created for technical objects, such as equipment or functional locations, facilitated by the Create Maintenance Request app. All essential data is input for screening, processing, planning, and execution of the request. Attachments and links can be included. Open requests are presented for review until submission.

  Prioritization is achieved by selecting from a priority list or evaluating priority based on consequence categories, consequences, and likelihoods. This assessment requires maintained prioritization profiles for a combination of maintenance plant and notification type. Upon request submission, the latest draft is accessible in the My Maintenance Requests app, categorized based on processing statuses.

- **Screening phase**: Submitted maintenance requests progress to the screening phase, where screening and acceptance occur. Supervisors review open maintenance requests through the Screen Maintenance Requests App. Inadequate information can prompt the return of the request to the initiator. Upon the provision of necessary details, supervisors review the request once more. The app groups maintenance requests according to their statuses.

The next phases involve planning, approval, preparation, and scheduling.

- **Planning phase**: Upon acceptance, a maintenance request transitions into the planning phase, becoming a maintenance notification. As a maintenance planner, you can create and plan orders. The planning phase includes creating orders requiring cost approval based on configuration. Workflow steps for approval are applicable based on the order type's configuration. Workflow approval can be automatic or manual, affecting the order's progression. Orders can be submitted for approval through various apps. The Preparation phase follows approval, enabling the division of maintenance effort, resource, and material planning.

- **Approval phase**: Maintenance order approval involves a flexible workflow process managed through the Manage Workflows for Maintenance Orders app. Workflow configuration includes step conditions, sequences, and approvers. Workflow steps with true conditions are executed, leading to order approval or rejection. Approved orders can be released, progressing to the Preparation phase.

- **Preparation phase**: Orders approved for execution enter the preparation phase, where resources, spare parts, and services are coordinated. This phase ensures efficient workload distribution and availability checks. Subsequently, orders proceed to the scheduling phase.

- **Scheduling phase**: Orders are dispatched for execution, signifying readiness. Resource Scheduling, available with an additional license, can be used to manage specific work centers. Material availability checks can be conducted through various apps.

The final phases are execution, post-execution, and completion.

- **Execution phase**: In this phase, maintenance technicians execute tasks. Preliminary (PRE) and main (MAIN) operations are typically performed in this phase. After task completion, the order's status is updated to Main Work Completed.

- **Post-execution phase**: Technicians conduct post-execution tasks, such as time recording, malfunction detail preparation, and confirmation. Supervisors review failure data and approve orders for technical completion, which leads to the Completion phase.

- **Completion phase**: This phase entails comprehensive review, financial settlement, and eventual business completion of the maintenance orders. This phase concludes the maintenance process, achieving a closed order status.

Phase Control Codes

Phase control codes are employed to regulate phase transitions. By activating a phase control code to block a specific phase in a maintenance order, the phase transition is disallowed until the corresponding code is deactivated. These codes can be activated for both order headers and operations.

4.12. Customizing Settings for a Phase-Based Maintenance Process

This section describes the Customizing settings for a phase-based maintenance process. Table 4-7 lists and describes IMG Settings for Phase-Based Maintenance Process.

Table 4-7. *IMG Settings for Phase-Based Maintenance Process*

Field Name and Data Type	Menu Path
Business Feature	ABAP Platform → Application Server → Business Management → SAP Business Feature → Activate Business Feature
Event Prioritization - Consequences and Likelihoods	Plant Maintenance and Customer Service → Maintenance and Service Processing → Maintenance and Service Notifications → Notification Processing → Response Time Monitoring → Event Prioritization → Define Consequences Categories, Consequences, and Likelihoods
Event Prioritization - Prioritization Profiles	Plant Maintenance and Customer Service → Maintenance and Service Processing → Maintenance and Service Notifications → Notification Processing → Response Time Monitoring → Event Prioritization → Define Prioritization Profiles

(continued)

Table 4-7. (*continued*)

Field Name and Data Type	Menu Path
Detection Methods	Plant Maintenance and Customer Service → Maintenance and Service Processing → Maintenance and Service Notifications → Notification Creation → Notification Content → Additional Functions for Notification Types → Define Detection Methods
Order Type	Plant Maintenance and Customer Service → Maintenance and Service Processing → Maintenance and Service Orders → Functions and Settings for Order Types → Configure Order Types
Template for the Material Availability Check	ABAP Platform → Application Server → System Administration → Activation of Scope Dependent Application Job Catalog
Activate Event Type Linkage for Procurement Milestones	Plant Maintenance and Customer Service → Maintenance and Service Processing → Maintenance and Service Orders → Functions and Settings for Order Types → Procurement → Activate Event Type Linkage for Procurement Milestones
Define Assignment Rules for Procurement Milestones	Plant Maintenance and Customer Service → Maintenance and Service Processing → Maintenance and Service Orders → Functions and Settings for Order Types → Procurement → Define Assignment Rules for Procurement Milestones
Overall Status Profile	Plant Maintenance and Customer Service → Maintenance and Service Processing → Fiori Apps for Maintenance Processing → General Settings → Configure Overall Status Profiles

(*continued*)

Table 4-7. (*continued*)

Field Name and Data Type	Menu Path
Phases and Subphases	Plant Maintenance and Customer Service → Maintenance and Service Processing → Fiori Apps for Maintenance Processing → General Settings → Define Subphases and Map to Overall Statuses

4.13. SAP GUI Maintenance Process

Your organization aims to introduce asset management using SAP S/4HANA and has chosen to collaborate with SAP GUI for the user interface, opting for it instead of web-based UIs. You want to understand the process of generating notifications and orders through the conventional SAP GUI user interface.

4.13.1. Performing Maintenance Tasks through SAP GUI

From a procedural standpoint, the sequential stages for planning and executing maintenance tasks in SAP GUI mirror those in web-based interfaces. However, SAP GUI transactions stand apart from web applications not only in terms of user interface but sometimes also in functionality.

Therefore, delving into the SAP GUI-centric process is valuable, especially for proficient users or application consultants primarily engaged in backend activities like master data creation or system configuration.

The following sections briefly outline the procedure for planning and executing a maintenance task via SAP GUI, occasionally highlighting specific functions exclusively accessible through SAP GUI.

4.13.2. Order Hierarchies

Maintenance orders can be organized in a hierarchical structure. Subsequent orders can be generated in relation to an existing maintenance order using a dedicated transaction. The initial order crafted becomes the apex node of the hierarchy and is referred to as the main order.

An order hierarchy can prove immensely useful when disparate work centers collaborate on the same maintenance task while necessitating separate cost considerations and processing.

To create a sub-order for an existing main order, you can employ transaction IW36.

4.13.3. Material Planning

Within the scope of a maintenance order, you can select spare parts directly from a 3D model utilizing the SAP 3D Visual Enterprise viewer. This process necessitates linking the appropriate document, categorized as document type SP.

SAP Visual Enterprise Author assigns valid material numbers to the model's spare parts.

When a valid SAP material number cannot be identified, it becomes unfeasible to associate the spare part with the spare part list.

Access to the graphical representation is possible through the Operations tab and the operation-specific details when utilizing SAP GUI transactions within the maintenance order. In the task list, the VE-Viewer button is accessible via the component overview, serving as a gateway to display the visual representation of the spare parts.

4.13.4. Visual Task Lists

Visual task lists depict sequential work processes founded on a document featuring an animated 3D file (RH file). This sequence of operational steps can be presented comprehensively through the Visual Enterprise Viewer or in a step-by-step manner.

Within maintenance orders and task lists, each operation can be linked to a visual task list (on either the operation screen or the operation-specific detail screen). Upon successfully linking a visual task list at the operation level to the maintenance order or task list, the task list's content can be played and viewed.

4.13.5. Material Availability Verification

When scheduling stock materials for order operations, the system can perform a comprehensive check to determine whether all required materials within the maintenance order are available in sufficient quantities. The Availability Check function fulfills this task.

Depending on the system configurations and the data entered in material master records, the system conducts an availability check for all materials linked to the operations in the maintenance order. The outcome of this check is communicated to you through an automated online message. In cases where materials are insufficiently available, the system generates an error log containing detailed information about the check's outcome.

Upon releasing a maintenance order, the system can execute an availability check for planned materials, subject to your customized settings. If the check reveals that certain planned materials lack enough, it's possible to proceed with order release based on your system's settings.

The material availability list furnishes insights into the availability of materials planned for an order.

- For non-stock items, this list informs whether the planned goods receipt (GR) date aligns with the operation's earliest or latest start date. It also provides reasons for any potential inability to meet the GR deadline.

- In the case of stock items, the quantity is determined and confirmed as available or unavailable through the material availability check.

In SAP S/4HANA, an additional transaction, IW38A, is available for performing a material availability check.

Note   The list presents valid material availability data based on available system records. It doesn't display simulation data used for calculating appropriate start dates.

You also have the option to mark specific order components as irrelevant for planning. This signifies that no reservation or purchase requisition is generated for such components.

The material availability list can be initiated within an order or from the order list editing function. If accessed from the list editing function, the list can be displayed at various summarization levels, including order level, orders with operations, and material items.

4.13.6. Capacity Planning

Capacity planning within the SAP system empowers you to regulate and supervise the capacity load across your various workshops, facilitating the synchronization of capacity supply with demand. The maintenance work center (workshop) defines the capacity supply in the master record, while the planned maintenance orders serve as indicators of capacity demand.

Capacity planning encompasses the following components.

- Capacity evaluation
- Capacity leveling

During capacity evaluation, the capacity requirements are compared to the available capacity, which is what is accessible per working day and is maintained within the maintenance work center.

Capacity requirement delineates the capacity demanded by orders at a specific point in time.

Capacity leveling aims to equalize exceeded and unutilized capacities of work centers, enabling optimal utilization of employees, machinery, and appropriate resource selection.

In addition to capacity planning for maintenance work centers (workshops), detailed planning at the individual level is achievable. When planning at this level, individuals can be scheduled and organized based on a graphical or tabular planning board, facilitated through integration with human capital management (HCM), which provides data like qualifications and attendance records.

4.13.7. Worker Safety

Worker safety encompasses various aspects.

- Permits

- Worker Safety (web functions)

- Integration with Risk Management System (GRC)

- Work Clearance Management (WCM)

4.13.8. Permits

Order release can also be contingent upon permits to adhere to health and safety regulations and oversee order processing. Permits can be automatically assigned to the order header based on predefined criteria, determining order release timing.

Technical permits, like welding permits, are manually assigned to technical objects and aren't classified. When an order is generated for the technical object, these permits are copied to the order and can influence release based on specific settings.

Process-oriented permits are automatically determined based on order header attributes, such as planned costs, potentially influencing order release. This automatic determination hinges on permit classification.

4.13.9. Work Clearance Management

Essential maintenance tasks on technical objects require implementing safety measures before execution, ensuring a secure working environment. These measures include lockout/tagout, fire protection, and radiation protection.

Work Clearance Management (PM-WCM) oversees these safety measures, ensuring safe working conditions for maintenance staff, compliance with environmental regulations, and technical system reliability.

4.13.10. External Services

In maintenance orders, external operations are initiated by assigning control key PM02. The description of the external service is integrated into the order operation as a long text. Planning external services generates a purchase requisition in the background, which the purchasing department converts into a purchase order.

External services aren't confirmed with time entries; instead, GR for the purchase order serves as confirmation. The service is assigned the value of the purchase price, entered in financial accounting. The debiting of the maintenance order with this value is recorded in financial accounting.

Within the maintenance order, the external processing screen for each operation indicates whether a GR has been posted for the purchase order (PO), displaying the posted GR quantity.

Vendor invoices typically follow deliveries, with offsetting entries recorded in a goods receipt/invoice receipt clearing account (GR/IR clearing account), automatically adjusted upon invoice receipt. Any discrepancies between PO and invoice values are adjusted in order.

4.13.11. Technical Confirmation

Detailed technical findings during maintenance activities form the basis for subsequent evaluations. These findings encompass details like the cause of damage, executed tasks, exact damage location, machine downtimes, and system availability.

Technical findings can be recorded in the malfunction report (if applicable) or an activity report post-order creation. Upon completing the maintenance notification, the system transfers data to the notification history, containing information about each technical object's damage, malfunctions, causes, findings, and performed maintenance work. The TECO status can be reversed as needed.

4.13.12. Document Flow

The document flow overviews all document types generated during order processing, including notifications, time confirmations, goods issue/receipt, purchase requisitions, purchase orders, quotations, service entry sheets, and invoices.

4.13.13. Action Log

The action log chronologically documents changes to notifications, orders, equipment, and functional locations. It offers insights into who altered data or statuses in various fields and when. Activating change document creation for relevant objects is a prerequisite for using this function.

4.14. Refurbishment Maintenance Process

Your company possesses valuable components that aren't viable for internal repair or restoration. As a result, you intend to dispatch them to a service company staffed by qualified experts for refurbishment. Concurrently, you aim to maintain constant visibility into the condition and status of these components. Hence, you need to comprehend the external refurbishment process for spare parts, as shown in Figure 4-14.

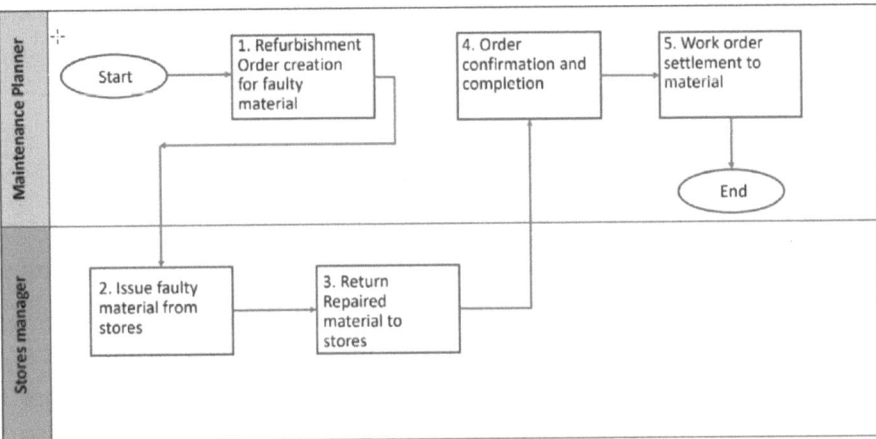

Figure 4-14. *Refurbishment process*

Refurbishing valuable components encompasses the following.

- Derived from a designated inventory

- Forwarded to an external subcontractor

- Initiated by a maintenance directive

- Interlinked with Material Requirements Planning (MRP) and procurement

The merits of refurbishment encompass the following.

- Enhanced identification of faulty components

- Closer integration of Materials Management and Plant Maintenance

- Distinct inventory for externally processed components

- Fusion of subcontracting with procurement

- Inclusion of serial numbers within purchase orders

4.14.1. External Refurbishment through Subcontracting

Subcontracting involves sending refurbishment parts to external subcontractors with or without serial numbers. A service provider obtains a part (also known as a *rotable*) from a client for upkeep or repair. Initially, the service provider incorporates the defective part into its inventory. Subsequently, it conducts internal repair or forwards the part to another subcontractor for refurbishment, repair, or maintenance. Upon completion of the work, the subcontractor returns the part to the service provider, who then returns it to the customer.

The genesis of the external refurbishment process can be traced to the rotables and subcontracting business process within the aerospace and defense industry solution. Various process variations are available. From a maintenance perspective, the external refurbishment process is activated through a refurbishment order.

The following explains the process of subcontracting with a direct order reference.

1. Initiation involves generating a maintenance order with subcontracting as the target, accompanied by a subcontracting task featuring the relevant part and material provision indicator.

2. The subcontracting task bears the characteristics of an external activity (control key PM02) with a subcontracting flag. This task leads to creating a purchase requisition upon order saving or release.

3. The PO is formulated in the purchasing department, referencing the purchase requisition from the maintenance order. The PO contains the serial number (the defective unit), any other necessary spare parts, and a specification of the expected state (batch) upon delivery.

4. The subcontracting monitor facilitates the creation of a delivery associated with the PO. Following this, a goods issue (GI) is recorded concerning the delivery. Once the GI is logged, the defective component is physically transferred to the service partner. This component is stored within a distinct inventory in the system (material provision to the vendor), ensuring its visibility during external refurbishment.

5. Repair is undertaken either by the service partner or its subcontractor.

6. The GR for the refurbished component is logged in the spare parts warehouse using a new valuation type.

7. The refurbished component is now ready for use. A maintenance order pertaining to subcontracting is marked as technically complete.

Subcontracting via MRP with Indirect Order Reference

Subcontracting facilitated by MRP is based on separate inventory management in different storage locations. In this scenario, faulty parts are housed in a dedicated inventory linked to their corresponding state. The storage location for defective parts is designated as irrelevant to MRP. The material master contains a subcontracting special procurement indicator. When regular stock is insufficient, MRP can opt for either conventional procurement or subcontracting.

In a subcontracting scenario, parts are sourced from the defective parts inventory.

1. The maintenance order serves as an indirect trigger for subcontracting. Although it lacks a direct subcontracting task, it outlines a routine maintenance activity associated with the spare part.

2. The reservation initiated by the maintenance order results in stock shortfall.

3. MRP's planning process generates a planned order. This order can be transformed into a purchase requisition/purchase order or an internal refurbishment order. The planned PO features a subcontracting item linked to the defective unit.

4. The subcontracting monitor aids in creating and selecting a delivery for the PO. A goods issue (GI) is then recorded for the delivery. Subsequently, the defective component is physically transferred to the service partner. This component remains in a specialized inventory within the system (material provision to the vendor), ensuring its visibility during the external refurbishment process.

5. The service partner or its subcontractor performs
    the repair.

6. The GR for the refurbished component is registered
    in the spare parts warehouse, utilizing a new
    valuation type.

7. The revitalized component is now available for use.
    The maintenance order related to subcontracting is
    marked as technically completed.

4.15. Preventive Maintenance Process

To minimize downtime and reduce maintenance expenses, it's imperative
to regularly inspect and uphold the technical systems within a company.
This underscores the importance of grasping the preventive maintenance
process (see Figure 4-15).

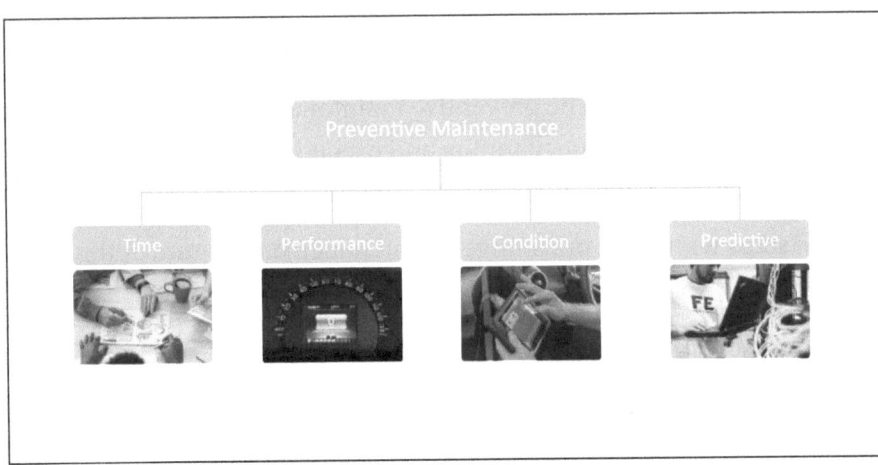

Figure 4-15. *Various types of preventive maintenance*

Preventive maintenance can be classified into the following four distinct categories.

- **Time-based**: Within time-based preventive maintenance, maintenance tasks are initiated following the passage of a specified duration. For instance, every six months.

- **Performance-based**: In performance-based preventive maintenance, maintenance tasks are initiated upon reaching a specific performance level (counter reading); for instance, after every 10,000 km.

- **Condition-based**: This type of preventive maintenance triggers maintenance tasks when a condition deviates from a predetermined value range. For example, when thread depth falls below 15 mm or temperature exceeds 85°C.

- **Predictive maintenance**: Also known as machine control, it is cloud-based and real-time, utilizing collected data to anticipate machine failures (IoT applications). The preventive maintenance process encompasses the planning and execution of periodic inspection and maintenance activities.

The steps within the preventive maintenance process are outlined as follows.

1. The task list delineates the process steps to be executed, which can either be reliant on the specific object or independent thereof.

2. A maintenance plan is generated for the object, automatically generating orders, notifications, and service entry sheets in alignment with designated guidelines.

3. Scheduling is responsible for regularly calling up orders, notifications, and service entry sheets while recalculating planned dates.

4. The maintenance order is automatically generated by scheduling the maintenance plan and is recorded within the order list for subsequent processing alongside other orders.

5. Technical completion marks the order and its corresponding planned date within the maintenance plan as concluded. This date of technical completion plays a role in calculating the next planned date within the maintenance plan. Further insights into preventive maintenance are covered in the upcoming unit.

4.16. Inspection Checklists

Checklists find frequent utility in Plant Maintenance for executing inspections and assessments of assets. These checklists also carry the weight of legally binding documentation.

The inspection checklist process represents a comprehensive cycle encompassing the creation of inspection plans, recording results, and follow-up steps within Plant Maintenance.

Given that inspections are conducted at regular intervals, maintenance orders are initiated based on predefined maintenance plans. The initiation of an inspection checklist linked to a maintenance order occurs under the following triggers.

- Classification data allocated to a technical object along with an inspection plan

- Checklist type associated with a maintenance order operation

Quality Management (QM) inspection lots are established and matched with the corresponding inspection checklists during the inspection checklist process.

By configuring the inspection checklist process within your system, you benefit from streamlined efforts in maintaining master data and executing operational tasks.

To facilitate a recurring checklist process, you must regularly establish the following master data components.

- PM Technical Object

- PM Work Center

- PM Maintenance Plan

- PM Task List

- QM Inspection Plan

Inspections are predicated on work orders, which are automatically generated through maintenance planning. Matching the inspection plan's classification characteristics occurs with the technical object classification and the maintenance order operation's checklist type (or with business add-ins).

- **Generating object lists**: Upon or after creating a Plant Maintenance (PM) work order, an object list can be created from the order's header object and objects from the maintenance plan item. Subsequently, during inspection checklist generation, all technical objects within the order's object list are examined for congruent inspection plans.

- **Manual generation in maintenance order transactions**: The Generate button on the Objects tab in transactions like IW31 or IW32 can manually create an object list based on a reference object.

After manually generating the object list, press the Generation Log button to display new or removed objects.

- **Automatic generation when saving a maintenance order**: When a maintenance order is generated using transaction IW31 and specific criteria (order type and plant) configured for the inspection checklist process are met, an object list is automatically created upon order save.

- **Automatic generation when scheduling a maintenance plan**: If a maintenance order originates from a maintenance plan and relevant criteria are met, an object list is generated automatically when the order is saved.

- **Generating inspection checklists and inspection lots**: Inspection checklists can be created manually or automatically. To generate inspection checklists, certain prerequisites must be met.

  - An object list must be generated, and the OLGE (object list generated) system status must be set at the order header level.

  - The classification characteristic of the inspection plan must align with the checklist type of the maintenance order operation.

  - The inspection plan and the technical object must share the same class name.

  - During checklist generation, the system scans for matching values of classification characteristics in the QM inspection plan and the technical object.

Based on the located inspection plan, new inspection lots are formed and presented on the Checklists tab. The inspection lot origin is set as 89 by default, with the material number of the inspection plan transferred to the inspection lot.

- **Inspection checklist results recording**: Transaction IW91 (Checklists: Result Overview) can be used to view inspection checklist results. The results can be displayed at different levels.

  - Inspection characteristics

  - Inspection lots

  - Order operations

  - Technical objects

  The ability to include or exclude deactivated checklists in the results list is provided, and navigation from the results list to corresponding display transactions is possible.

- **Closing inspection checklists and maintenance orders**: Using the results list in transaction IW91, navigation to transactions QA13 (Display Usage Decision) and QA11 (Record Usage Decision) can be done to view or record the usage decision for an inspection lot. This is achieved by selecting the glasses icon or pen icon in the Usage Decision column.

- Transaction IW93 (Checklists: Collective Usage Decisions) enables collective usage decisions, along with follow-up actions like creating measurement documents.

In closing, setting a usage decision triggers follow-up actions defined in Customizing for the selected maintenance orders and inspection lots.

4.17. Mobile Maintenance

In SAP S/4HANA EAM, Asset Manager and Work Manager are mobile applications designed to enhance asset management and work execution processes. These applications enable users to perform various tasks, inspections, and maintenance activities while being mobile, thereby increasing efficiency and effectiveness. However, there are differences in terms of their focus and functionalities.

4.17.1. SAP Asset Manager

SAP Asset Manager is a next-generation, asset-centric mobile application that is tightly integrated with the SAP S/4HANA digital core and SAP Business Technology Platform. It is available for both iOS and Android platforms. The primary focus of SAP Asset Manager is on managing and maintaining assets in a more efficient and streamlined manner. It offers features that allow users to do the following.

- View and manage technical objects/assets

- Perform inspections, readings, and measurements related to assets

- Create and manage work orders for maintenance, repairs, and service tasks

- Handle notifications for unexpected events or issues related to assets

- Record time spent on different tasks

- Manage spare parts and components needed for maintenance tasks

- Utilize maps for geolocation and tracking

- Access and view relevant documents and attachments

SAP Asset Manager provides a user-friendly interface with integration to native device features, offering a holistic solution for managing assets, maintenance, and related processes on the go.

The following are the prerequisites to implement SAP Asset Manager.

- SAP S/4HANA on-premises 1610 FPS01 or higher

- Implementation of SAP S/4HANA Asset Management

- SAP Business Technology Platform

- macOS development environment

- iPads for deployment of the mobile app

4.17.2. SAP Work Manager

SAP Work Manager is an application for mobile asset management that evolved from the former Syclo Work Manager. It is also available for both on-premise and cloud editions. While SAP Asset Manager focuses on a broader range of asset-related activities, SAP Work Manager specifically targets work order management and execution. Users can utilize SAP Work Manager to do the following.

- Receive, manage, and execute work orders

- View detailed task lists and instructions for work orders

- Capture time and labor details

- Record completed tasks and update task statuses

- Request and manage spare parts and materials for work orders

- Access relevant documents and attachments for work orders

- Perform follow-up actions on work orders

4.17.3. Key Differences

The main difference between SAP Asset Manager and SAP Work Manager lies in their focus and scope of functionality.

- SAP Asset Manager offers a broader range of capabilities, encompassing asset management, inspections, readings, maintenance, notifications, and more. It's suitable for users who need to manage various aspects of assets beyond just executing work orders.

- SAP Work Manager is more narrowly focused on work order management and execution. It caters specifically to users responsible for carrying out tasks and activities outlined in work orders, such as maintenance technicians, field workers, and service personnel.

In essence, while SAP Asset Manager covers a wider array of asset-related activities, SAP Work Manager specifically addresses the needs of workers involved in executing work orders and related tasks. Organizations can choose the application that best aligns with their specific requirements and the roles of their mobile workforce.

4.18. Summary

This chapter explored various maintenance processes within SAP S/4HANA EAM. You gained an overview of the standard maintenance process, which establishes a fundamental framework for maintenance activities, and insight into a simplified maintenance process, offering streamlined approaches for swift task execution.

The phase-based maintenance process was introduced, illustrating how tasks are segmented into phases, aligning with the specific needs of each project phase. Transitioning to the SAP GUI maintenance process, the integration of graphical user interfaces was examined, enhancing user experience.

Next, the refurbishment maintenance process was covered, showcasing the process of rejuvenating valuable components through external services and you learned about the significance of preventive maintenance process, elucidating strategies to anticipate and manage asset issues proactively.

The chapter progressed to explore inspection checklists, highlighting how these comprehensive lists streamline inspection and recording activities. Concluding the chapter, the integration of was unveiled, enabling on-the-go task execution and data management through dedicated mobile applications. Collectively, these diverse processes cater to a spectrum of maintenance needs within SAP S/4HANA EAM, optimizing efficiency and effectiveness.

CHAPTER 5

Preventive Maintenance

This chapter delves into the essential aspects of keeping your assets in top-notch condition while harnessing the power of SAP S/4HANA. We explore a range of functionalities that cover everything you need to know.

The following are some of the key topics covered in this chapter.

- Preventive maintenance

- Master data

- Single-cycle maintenance plan

- Planning regular external service procurement

- Maintenance planning with a time-based strategy

- Maintenance planning with a performance-based strategy

- Maintenance planning with cycles of different dimensions

- User interfaces

© Rajesh Ojha and Chandan Mohan Jaiswal 2023
R. Ojha and C. M. Jaiswal, *SAP S/4HANA Asset Management*,
https://doi.org/10.1007/978-1-4842-9870-1_5

5.1. What Is Preventive Maintenance?

Preventive maintenance, or proactive maintenance, is a way of taking care of things before they break or stop working. It involves doing regular checks and minor repairs to keep things running smoothly and avoid bigger problems later on. Just like getting regular check-ups at the doctor helps keep you healthy, preventive maintenance helps keep machines, equipment, and facilities in good shape and prevents unexpected breakdowns.

All planned maintenance (preventive / proactive maintenance) can be grouped into the four types shown in Figure 5-1.

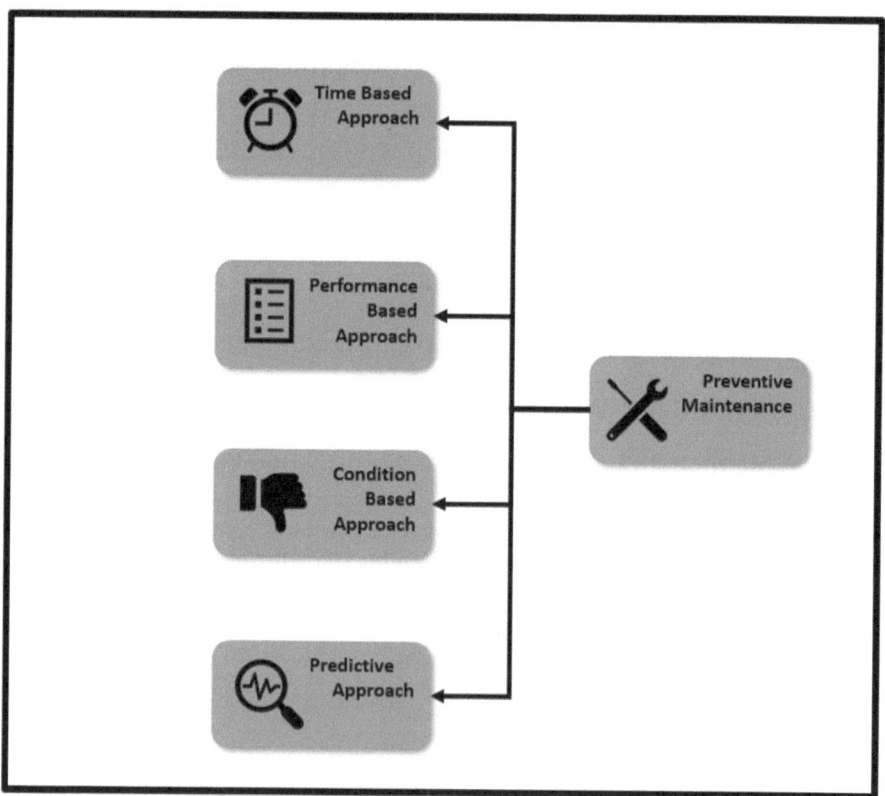

Figure 5-1. *Different ways to schedule preventive maintenance*

- **Time-based maintenance**: Maintenance tasks are scheduled at specific time intervals; for example, the air filter is replaced every three months.

- **Performance-based maintenance**: Maintenance tasks are scheduled when a specific number of operating hours or a certain reading is reached; for example, the oil should be replaced after 500 hours of engine running.

- **Condition-based maintenance**: Maintenance tasks are scheduled when the asset's or its parts' condition is outside the acceptable value range; for example, replacing a vehicle's battery is performed when the power level is below 25%.

- **Predictive maintenance**: Maintenance tasks are scheduled when real-time condition data of the asset or its parts signals a possibility of part failure; for example, if a machine's oil flow pressure sensor reading is very high, it indicates that the oil filter is clogged.

S/4HANA Asset Management includes detailed preventive maintenance functionalities that help companies care for their machines, equipment, and facilities. The functionalities help them schedule regular inspections, servicing, cleanings, and minor repairs. This way, they can avoid unexpected breakdowns that could disrupt their work and cost a lot to fix.

5.2. Master Data

Master data is akin to the foundation of a software application. It constitutes the fundamental information that a software application, such as SAP S/4HANA, employs to interact with business process data. Envision it as the crucial details regarding entities such as equipment, materials, customers, vendors, and more. For instance, if you're utilizing S/4HANA to oversee a manufacturing factory, the master data would encompass information about all the machines, their specifications, their locations, a list of tasks to be performed, and other indispensable particulars.

5.2.1. Maintenance Task List

The maintenance task list is among the most crucial master data sets for conducting planned maintenance and repair activities, including preventive maintenance, periodic maintenance, regular checks, and inspections. For example, as preventive maintenance, a specific maintenance job must be performed for all piston pumps at a regular frequency. This job consists of operations and replacement parts from wear and tear if needed (e.g., disconnection from the electrical power supply, closing incoming fluid to the pump, visual inspection for leakage, seal replacement, etc.).

This section delves into intricate details about the various types of maintenance task lists, their utilization in maintenance orders, the planning of routine inspections, and important customization settings.

Types of Task Lists

In S/4HANA Asset Management, there are three types of maintenance task lists.

- An **equipment task list (E)** is created for a specific piece of technical object. This task list is assigned to a particular equipment master, such as the task list for half-yearly servicing of an equipment master number 200000235.

- A **functional location task list (T)** is also created for a specific piece of technical object. This task list is assigned to a particular functional location master, such as the task list for a yearly overhaul of the engine assembly line represented as functional location master number 1000-EA-AL01.

- A **general maintenance task list (A)** is not linked to any specific technical object, such as an equipment master or functional location. It is created to be used for a group of technical objects which have similar technical characteristics. For example, a task list for quarterly preventive maintenance of a particular make of piston pumps with the same technical specifications.

Any of the task list types can be used for planned maintenance (e.g., preventive maintenance, routine inspection) and unplanned maintenance (corrective maintenance, breakdown maintenance).

Maintenance task lists are organized into clusters known as task list groups. Each task list group comprises maintenance task lists sharing identical or akin maintenance steps. The task lists in a task list group are recognized by the group counter, which sequentially labels the task lists within that specific group (see Figure 5-2).

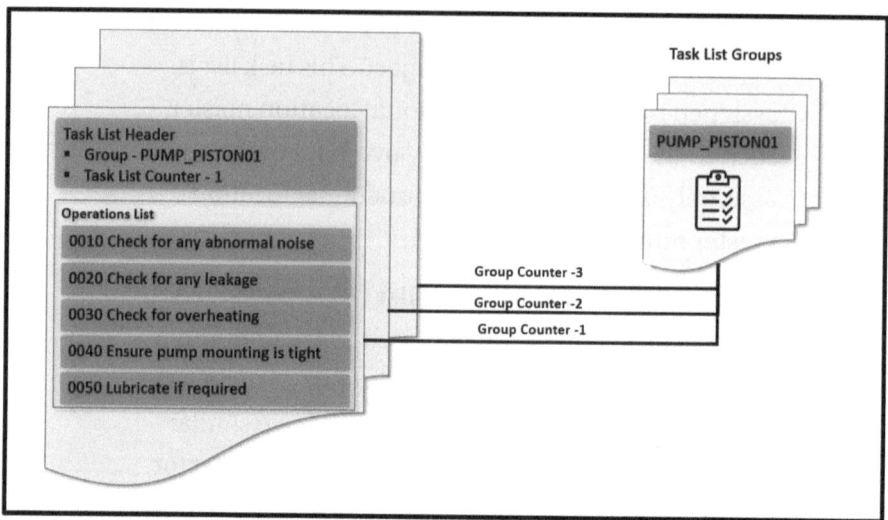

Figure 5-2. *Organization of maintenance task list in group counter*

Based on task list type, task list groups are either technical object-based (see Table 5-1) (functional location and equipment master task lists) or object-independent (see Table 5-2) (general maintenance task lists).

Table 5-1. *Equipment Task List Group*

Task List Group	Technical Object	Task List Counter	Description
3003	Equipment–200000235	1	3-month general service
3003	Equipment–200000235	2	6-month air filter replacement

Table 5-2. *General Maintenance Task List Group*

Task List Group	Task List Counter	Description
PUMP_PISTON01	1	Task list for pump general service
PUMP_PISTON01	2	Task list for pump piston replacement
PUMP_PISTON01	3	Task list for pump replacement

All the maintenance task lists within a group are managed as a single unit. Therefore, SAP suggests dividing your maintenance task lists into several smaller groups to streamline processing. This reduces the data volume that the system needs to handle when accessing a task list group, leading to shorter system response times.

Structure of Maintenance Task List

In general, data within a maintenance task list is divided into two sections called task list general data (referred to as header data in the SAP GUI user interface) and task list operation data. The header section data applies to the entire task list, whereas operation data is specific to each operation in the task list (see Figure 5-3).

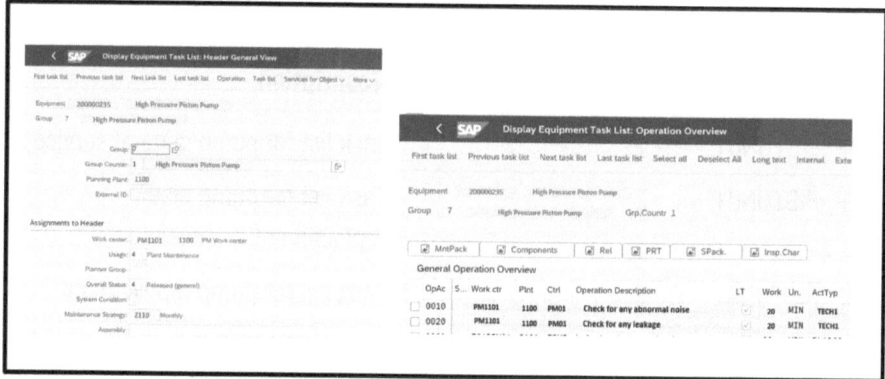

Figure 5-3. *Important data in header level and operation level of a task list*

The following are important pieces of information maintained at general data level.

- Plant

- Maintenance work center

- Planner group

- Maintenance strategy

- Assembly

- Quality management data

- Status

The following are some important data maintained at the operational level.

- Operation number

- Operation short description

- Maintenance work center

- Components required

- Production resource tool (PRT)

- Service packages

- Maintenance packages

Profile for Managing Task List

Specific fields in different maintenance task lists such as operation number and unit of duration typically contain the same data or information. To facilitate the user, this data can be defaulted during the creation of maintenance task lists using profile functionality. A profile holds standardized information needed in identical or similar combinations while processing maintenance task lists. Profile is defined in customization.

At the time of maintenance task list creation, you can select the previously defined profile, which contains the required data, on the initial screen. This data is automatically populated in the new maintenance task list. Users can overwrite the populated data at any time. Users can even default a particular profile ID (in case there are more than one profiles created) by using the User Parameter ID–PIN and the specific profile ID in the SAP user master data.

Operation Control Key

The control key functionality defines if an operation is processed internally (using a maintenance work center) or externally (using an external vendor) (see Figure 5-4). The control key also defines a few other processing behaviors of an operation, such as if the operation should be scheduled, relevant for costing, operation required to be confirmed, and printing allowed. Control keys are defined in customization.

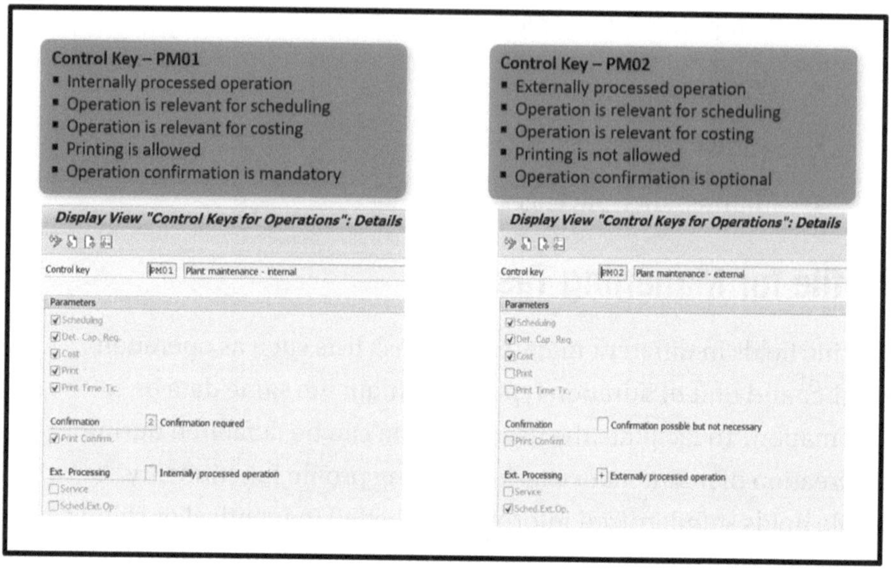

Figure 5-4. *Task list operation's control keys*

Adding Component in Maintenance Task List

In maintenance task lists, you can allocate material components to operations. These materials can be located and obtained from the bill of materials (BOM) associated with the technical object (equipment, functional location, or assembly) assigned to the task list. In this scenario, the BOM directly matches the content of the structure list.

Additionally, you can allocate stock materials, even if they aren't included in the technical object's BOM, directly to the operations within the maintenance task list. This is referred to as a "free material assignment." To accomplish this, you use the material number for the assignment. To perform a free material assignment, you need to define a BOM usage (typically in Asset Management) within the Customizing settings. When you opt for a free assignment, the system generates an internal BOM, which cannot be manipulated through the application. The material components that are assigned to the operations of the task list are copied into the maintenance order when the maintenance task list is transferred into the maintenance order.

Adding Component in General Maintenance Task List

To include material components in a general maintenance task list, you must begin by assigning an assembly (material master number of the assembly) in the header section of the general maintenance task list. It is possible to append components from the assembly's BOM to the operations within the general maintenance task list (see Figure 5-5). Additionally, you can incorporate material components from the common list of materials.

If you intend to modify or remove the assembly from the header section of the general maintenance task list, you can only do so after you have deleted component assignment from the operations.

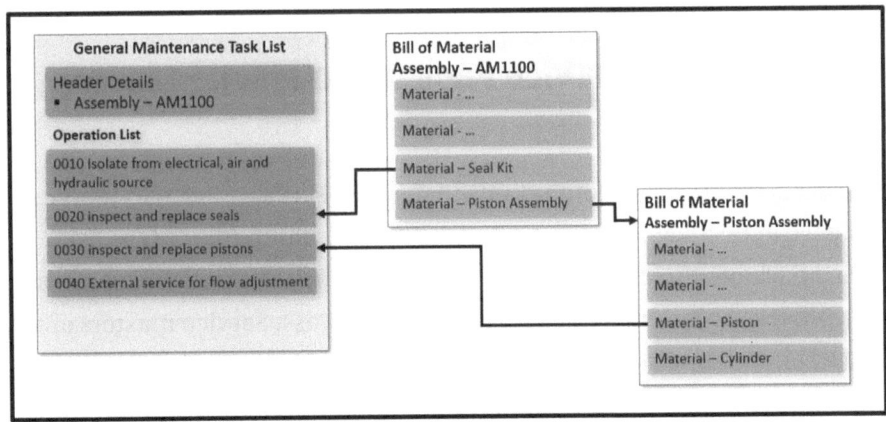

Figure 5-5. *Adding components in general maintenance task list*

Adding Component in Equipment Task List and Functional Location Task List

You can add material components in the equipment task list from the BOM for the equipment master and in the functional location task list from the BOM for the functional location. The material components you wish to allocate to a task list for equipment or functional location do not necessarily need to be present in the relevant BOM at the time of assignment. It is also possible to assign stock materials without restrictions. The system does not automatically include the freely assigned material in the BOM. The original form of the initial BOM is maintained, allowing you to access it whenever needed.

For the unrestricted allocation of materials in maintenance task lists, the system administrator needs to define a BOM usage for maintenance related BOMs (e.g., usage 4 in the SAP standard system) in the customizing configuration for maintenance task lists. After material has been freely assigned, avoid altering the designated usage, as modifying it could result in the potential loss of existing free material assignments.

Adding External Service Package to Maintenance Task List

For externally processed operations in task list, service packages (service master) can be assigned to the operation in the maintenance task list. A service package can be assigned to those operations as well which need to be processed internally in intercompany process. Service masters are created and maintained in Materials Management application. You have the option to assign service packages to the operation by either inputting a specific service number or choosing one or multiple services from the standard or model service specifications.

Apart from employing service packages, services can also be manually added within the service specifications for the operation. Nonetheless, all particulars, including price, unit of measure, and description, must be input manually.

5.2.2. Creating Maintenance Order with a Task List

General maintenance task lists are utilized in maintenance plans to facilitate the automatic copying of task lists into preventive maintenance orders and regular inspection orders. This copying occurs when orders are generated from these maintenance plans. Nevertheless, task lists can also be manually chosen during the manual creation of a maintenance order. For instance, if a user intends to execute ad hoc preventive maintenance for equipment lacking a maintenance plan, and a general maintenance task list applicable to comparable equipment types exists, it can be selected (see Figure 5-6).

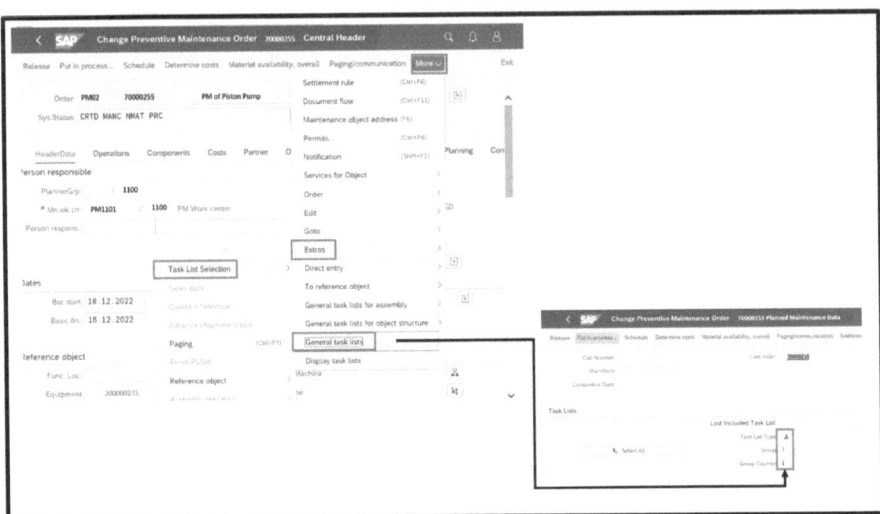

Figure 5-6. *Selecting the maintenance task list manually during the creation of a maintenance order*

5.2.3. Task List for Inspection Rounds

Inspection planning comprises similar repetitive inspections and tasks for many different technical objects. Users refer to their shop papers to perform inspection activities for specific technical objects. After completing the inspection activity, the maintenance technician confirms the manhours utilized and creates measurement documents in the system (see Figure 5-7). To set up inspection rounds, PRTs are required to be assigned as measuring points in the task list.

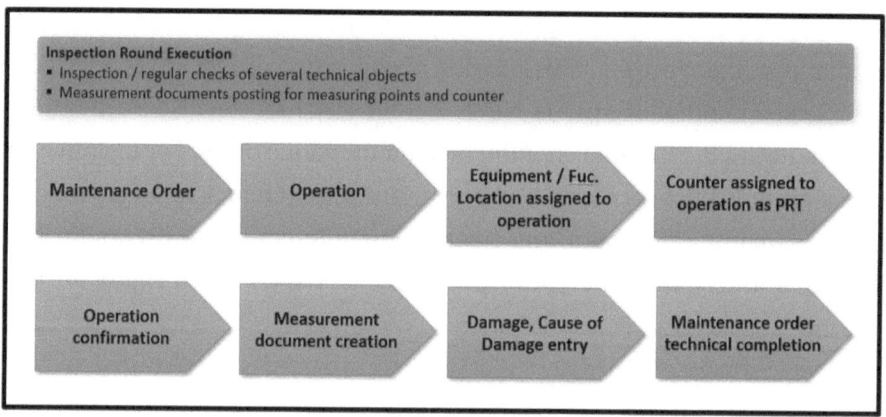

Figure 5-7. *Inspection round process flow*

The enhancement has been applied to the procedure for confirming overall completion. When conducting an overall completion confirmation at the conclusion of an inspection round, it is now possible to directly generate a notification from a technical object linked to an operation. Additionally, modifications can be made to an existing notification associated with a technical object.

You must activate the LOG_EAM_CI_3 and LOG_EAM_CI_4 business functions to have the full functionality of inspection rounds.

5.2.4. SAP GUI-Based (User Interface) Functions

This section details few of the important functionalities available from SAP GUI user interface.

Where-Used Lists

For the efficient planning of resources such as work centers, materials, and PRTs, for maintenance and repair work, the maintenance planner can utilize the "Where-Used Lists" report. This report allows the viewing of all maintenance task lists containing specific resources. For instance, if a material component becomes unavailable from a supplier and a replacement with an alternative component from task list operations where the discontinued material was assigned is needed, this report facilitates the process. Through this report, it becomes possible to substitute existing work centers and PRTs with alternative ones that are assigned to task lists.

Object Overview

Using the object overview report, you can display different objects that can be assigned to a maintenance task list. You can customize the list of objects to be displayed in the report. The report displays the following objects.

- Material
- Service package
- Object dependency
- Maintenance package
- Inspection characteristics

Action Log for Task Lists

Action log functionality records changes to maintenance task lists. The change log captures information such as who has made change, which field, old and new data and when the change happened.

5.2.5. Customizing Maintenance Task Lists

The following lists important objects that can be customized.

- Maintenance planner group

- Number ranges

- Profile for defaulting values in a task list

- Control key for operation

- Presetting for free assignment of material

- Presetting for multi-level list display

- Order type presetting for task lists

From the SAP Easy Access menu, navigate to Tools → Customizing. Double-click IMG → SPRO–Execute Project. Click the SAP Reference IMG button.

Table 5-3 lists important configuration paths related to plant.

Table 5-3. *Customizing Maintenance Task Lists*

Configuration Step	Configuration Path
Configure Planner Group	Plant Maintenance and Customer Service → Maintenance Plans, Work Centers, Task Lists and PRTs → Task Lists → General Data → Configure Planner Group
Define Number Ranges	Plant Maintenance and Customer Service → Maintenance Plans, Work Centers, Task Lists and PRTs → Task Lists → Control Data → Define Number Ranges for General Maintenance Task Lists
Define Profile with Default Values	Plant Maintenance and Customer Service → Maintenance Plans, Work Centers, Task Lists and PRTs → Task Lists → Control Data → Define Profiles with Default Values
Maintain Control Keys	Plant Maintenance and Customer Service → Maintenance Plans, Work Centers, Task Lists and PRTs → Task Lists → Operation Data → Maintain Control Keys
Presetting for free assignment of Material	Plant Maintenance and Customer Service → Maintenance Plans, Work Centers, Task Lists and PRTs → Task Lists → Control Data → Define Presetting for Free Assignment of Material
Presetting for multi-level list display (object overview)	Plant Maintenance and Customer Service → Maintenance Plans, Work Centers, Task Lists and PRTs → Task Lists → Presetting for List Display of Multi-Level Task Lists
Order Type Presettings for Task Lists	Plant Maintenance and Customer Service → Maintenance and Service Processing → Maintenance and Service Orders → Functions and Settings for Order Types → Default Values for Task List Data and Profile Assignments

5.3. Single-Cycle Maintenance Plan

S/4HANA Asset Management comes with detailed preventive maintenance functionalities, which help companies take care of their machines, equipment, and facilities. It helps them schedule regular inspections, servicing, cleanings, and minor repairs.

5.3.1. Maintenance Plan

The maintenance plan serves the purpose of planning, scheduling, and creating call objects for technical objects, such as equipment and functional locations, for designated dates. These call objects encompass maintenance notifications, orders, and service entry sheets (see Figure 5-8). Additionally, the option exists to generate maintenance notifications and orders simultaneously from the maintenance plan.

For example, for a three-month general servicing of air-conditioning, a job order must be created automatically regularly.

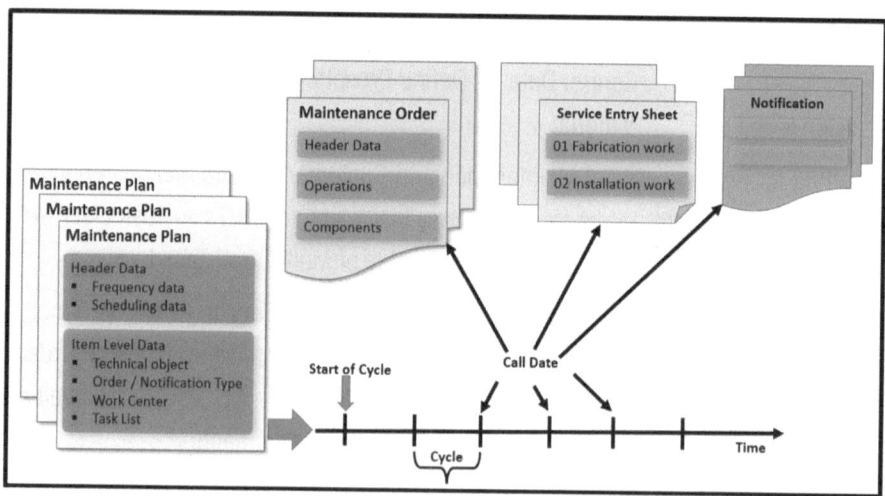

Figure 5-8. *Maintenance plan and types of call objects*

A maintenance plan can be scheduled according to duration or performance value. For instance, an air conditioner might require servicing every three months, while a car may need maintenance every 5000 kilometers. To meet these needs, users must generate two maintenance plans: one based on time and another based on performance.

Further, a maintenance plan can be scheduled based on a single cycle, multiple cycles (strategy based) or cycles of different dimensions (the multiple counter plan).

Single-Cycle Plan

In a single cycle maintenance plan, one schedule (one frequency) is adhered to for producing call objects like maintenance orders. Each time, identical operations and components are entered within the maintenance order.

Strategy Based

A strategy-based maintenance plan follows more than one schedule (multiple frequencies) to generate call objects such as maintenance orders. Maintenance orders generated for different schedules may contain varying operations and components. For example, servicing an air conditioner every three months requires inspection and cleaning, whereas servicing it every twelve months may necessitate the replacement of filters or gas refilling.

Multiple Counter Plan

A plan with multiple counters is scheduled using cycles of various dimensions (see Figure 5-9). This plan generates call objects, such as maintenance orders, when any single cycle reaches the designated limit or all cycles simultaneously hit the limit. For instance, a car needs servicing every six months or after covering 9000 kilometers, whichever occurs first. Conversely, a bus requires servicing when it accumulates 15,000 kilometers, and has been six months since its last maintenance.

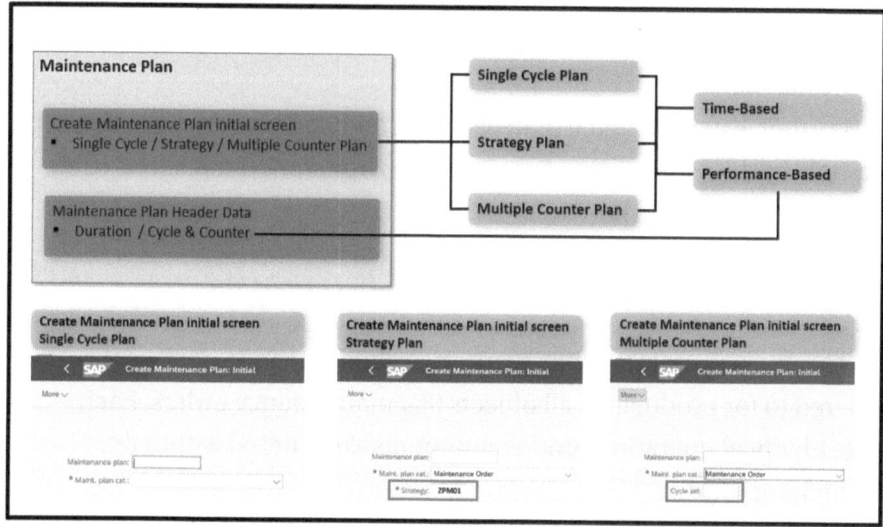

Figure 5-9. *Time-based and performance-based scheduling options for maintenance plan*

Structure of a Maintenance Plan

A maintenance plan consists of mainly two sections of data: maintenance plan header data and item data.

Information in the header data area applies to the entire maintenance plan. The following lists various data maintained at the header level.

- Maintenance plan number

- Maintenance plan text

- Cycle data: cycles, measurement units

- Scheduling parameters: call horizons, scheduling periods, tolerance

- Additional data: category, key date

Information in the item data area applies to a specific item (such as particular equipment or functional location) (see Figure 5-10). The following are various data maintained at the item level.

- Maintenance item number

- Technical object: equipment or functional location

- Planning data: Planning plant, call object type, planner group, task list

A maintenance plan can consist of several items (such as multiple equipment or functional locations). A separate call object (order, notification, or service entry sheet) is created for each item.

As an example, within a maintenance plan, diverse components of a water pump need to be managed. You can establish a maintenance item for the pump unit, a separate item for the electric motor, and a third one for the pump gears. Each maintenance item is linked to its individual task list. All these items fall under the umbrella of a single maintenance plan and share identical timings (scheduling data).

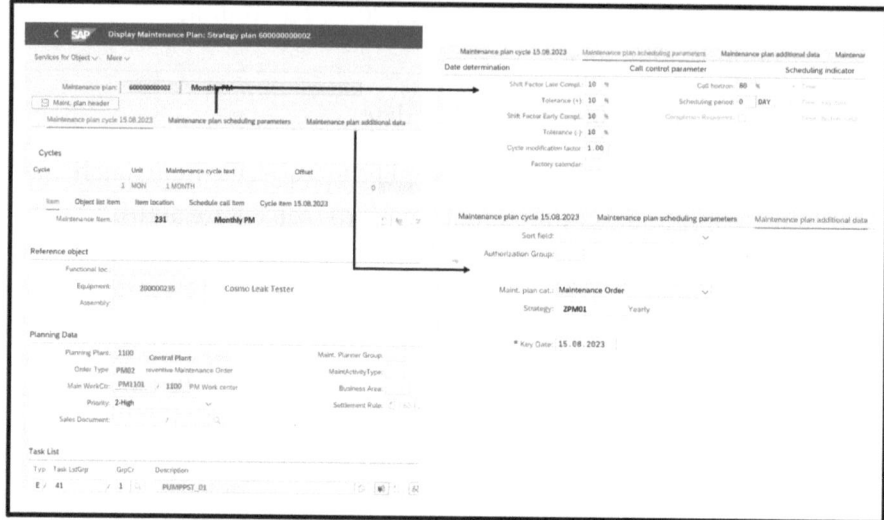

Figure 5-10. *Structure of a maintenance plan*

5.3.2. Maintenance Plan Setup for Generating a Maintenance Order

For internal planned maintenance (including preventive maintenance and routine inspections) of a technical object, maintenance orders are produced from the maintenance plan based on the frequency recommended by the manufacturer or the organization's maintenance procedure. The maintenance plan automatically generates these orders (see Figure 5-11).

Within the item data section of the maintenance plan, users can input the technical object (equipment or functional location) and the maintenance order type under which the order is generated. By selecting the Do Not Release Immediately checkbox, the maintenance order is not immediately released upon creation from the maintenance plan, even if the Release Immediately checkbox is set in customization for the order type.

Within the item data section of the maintenance plan, the task list is allocated. This task list encompasses the operations to be carried out and components to be replaced for the technical object. If there is no appropriate task list accessible for the technical objects, users have the option to directly create a new task list (general maintenance/equipment/functional location task list) from the maintenance plan. When no task list has been assigned in the maintenance item section of the maintenance plan, the system duplicates the maintenance plan text and uses it as the description for the first operation in the generated maintenance order.

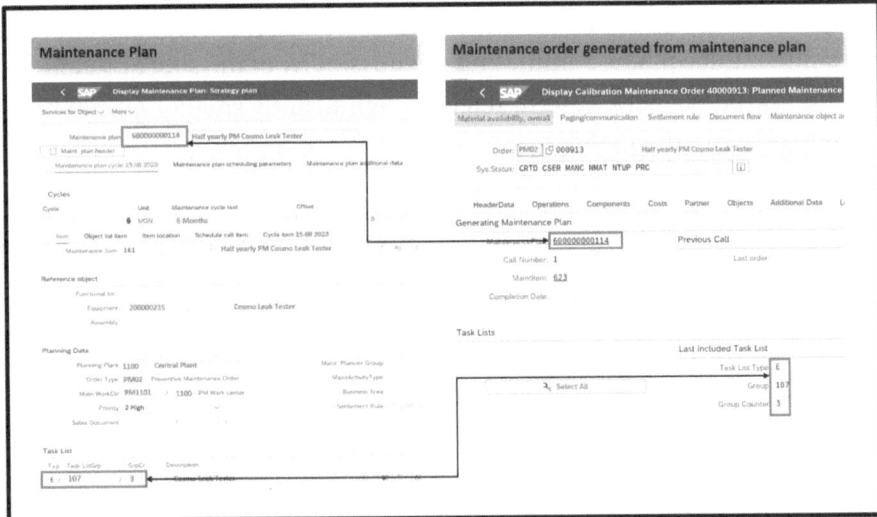

Figure 5-11. *Maintenance plan and autogenerated maintenance order*

5.3.3. Scheduling a Single-Cycle Plan

Once a maintenance plan is created, it is activated through the scheduling features of preventive maintenance. Maintenance plan scheduling is employed to produce call objects (maintenance orders/notifications) punctually.

The existing transaction code, IP30, has been employed to manage the scheduling of maintenance plans. Since processing all maintenance plans in IP30 can be time-consuming, different selection parameters are utilized to reduce the volume of maintenance plans during the scheduling process. However, as time progresses, new safety and legal regulations recommend timely creation of all call objects. Introducing the new transaction code, IP30H, now enables scheduling of all maintenance plans within a specified timeframe at once. When using transaction code IP30H, there is no need to input selection parameter values, as the system internally carries out preselection based on factors like maintenance strategies, dates, and counter readings. This ensures that the system incorporates only the due maintenance plans, eliminating the possibility of missing any call objects. With IP30H, the system selects a significantly smaller number of maintenance plans, leading to improved response times.

Scheduling Options

The following describes the scheduling options.

- **Start**: The Start option is for the first-time scheduling of a maintenance plan (see Figure 5-12). If during creation of the maintenance plan, start date was entered in the scheduling parameters section of the maintenance plan then the date is proposed during starting the scheduling else, user is required to enter start date.

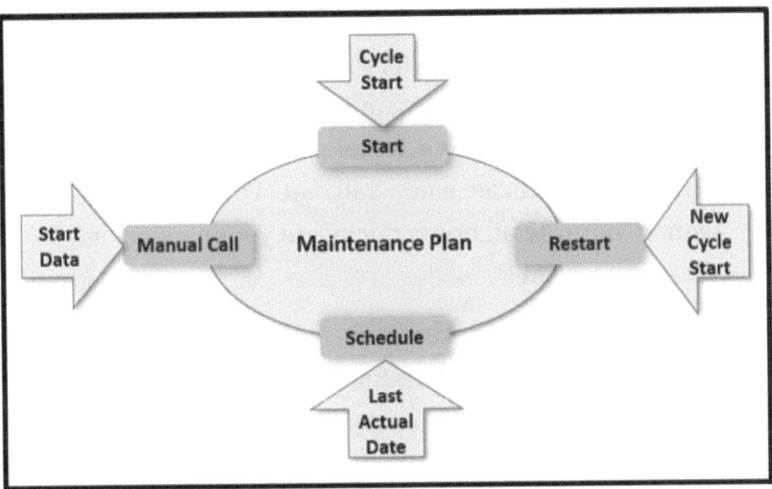

Figure 5-12. *Scheduling options in single-cycle plan*

- **Restart**: If the incorrect start date is mistakenly chosen or there have been alterations to the framework conditions, it's possible to schedule the maintenance plan anew. Any existing scheduled plan dates that are pending can be deleted. Deleting these pending plan dates does not impact calls that have already been executed.

- **Schedule**: It computes fresh planned dates and call dates, and (when relevant) triggers the subsequent maintenance order. Scheduling can be carried out on an individual plan basis manually, or collectively through online deadline monitoring or a system job. The method of collective scheduling is the most frequently employed.

- **Manual Call**: If you wish to schedule a maintenance plan, call for a specific date, you have the option to do so manually. You can use a manual call to include extra dates without disrupting the regular scheduling. To achieve this, indicate a new call date. For example, you want to perform ad hoc servicing on your car during the rainy season.

- **Deactivate**: This feature allows you to temporarily restrict maintenance plans for a specific duration. Subsequently, the system assigns the status INAK to the maintenance plan. This status prevents scheduling, and any pending planned scheduled dates are marked as blocked. Initiating maintenance calls is not feasible in this state. You can deactivate a maintenance plan in both the Change and Scheduling modes. The reactivation of blocked maintenance plans is possible whenever needed. For example, equipment has been sent to an external vendor for a major overhaul. During the period when the equipment is unavailable, the maintenance plan needs to be deactivated to cease the generation of maintenance orders.

Planned Date and Cycle Start Date

In the single-cycle plan, a value for the cycle (such as three months or one year) is assigned. The cycle value is intended for defining the frequency (interval period) for the planned dates. Call objects (orders/notifications) are generated based on the planned dates and the call horizon maintained in the scheduling parameters area of the maintenance plan. The call date (the order creation date) is typically set before the planned date, allowing enough buffer time for maintenance work planning, such as arranging external services or procuring non-stock materials.

The cycle start specifies the date at which the calculation of the planned dates should commence (see Figure 5-13).

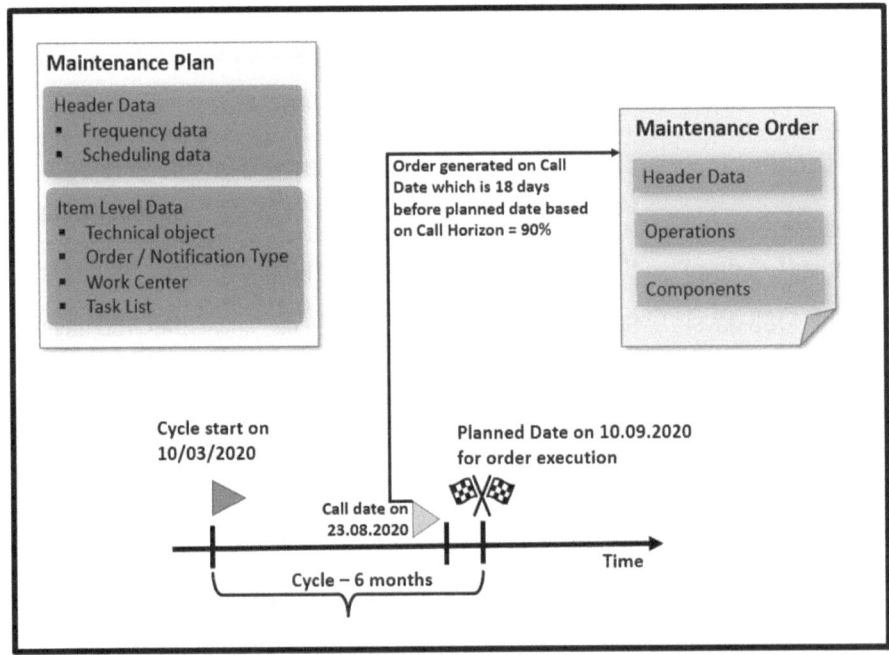

Figure 5-13. *Cycle start date, call date, and planned date*

Call Horizon

The call horizon is an important scheduling parameter used to calculate the call date, which is when a maintenance order, notification, or service entry sheet is created before the planned or actual execution date. The precise management of order generation allows you to plan the order ahead of the order execution date. Consequently, tasks that must be executed and finished on time can be achieved on the intended date. In a performance-based maintenance plan, it is recommended to always use the call horizon.

Call horizon is maintained as a percentage of the total cycle value. A 100% call horizon means that the call object is created on the planned date. 0% call horizon means that the call object is created on the date when the maintenance plan is started (see Table 5-4).

Table 5-4. *Call Horizon Calculation Example*

Cycle Length	Call Horizon	Call Date (order generation date)	Planned Date (execution date)
6 months (180 days)	Blank	Immediate (0% of 180 days)	after 180 days
6 months (180 days)	90%	after 162 days (90% of 180 days)	after 180 days
6 months (180 days)	100%	after 180 days (100% of 180 days)	after 180 days

You can maintain the call horizon in various objects of the preventive maintenance application, such as in a single-cycle plan, maintenance strategy, and strategy plan. Changes in the call horizon maintained within the maintenance strategy does not have a retrospective impact on existing maintenance plans using the maintenance strategy. The changed value is proposed in all maintenance plans created after the call horizon change.

When the LOG_EAM_CI_6 business function is activated, the call horizon value can be entered as days. You can maintain call horizon in percentage, number of calendar days or number of working days. The call horizon determines the timing for generating a maintenance call object, such as a maintenance order. It indicates the duration required between the order creation date and the upcoming planned maintenance date Should you indicate a percentage, the system compute the timing for the maintenance call using this specified percentage of the maintenance cycle.

If you opt for days (DAY), the system produces the maintenance call objects a specific count of days prior to the planned date. During the scheduling of the maintenance call, the system does not consider weekends, holidays, or any vacation shutdowns your company might have.

If a factory calendar is entered in the header section of the maintenance plan, the system uses the calendar for calculating dates during scheduling. In the case where a factory calendar is not maintained at the header level, the system retrieve the factory calendar of the planning plant that is maintained at the item level of the maintenance plan. For each item in the plan, the system calculates the earliest available working day as the planned maintenance day.

The call horizon functionality is not applicable for a multiple counter plan. In the scheduling parameters section of a multiple counter plan, you can enter a preliminary buffer in the number of days, which indicates how many days before the planned maintenance date the order should be created.

Scheduling Period

The scheduling period is one of the control values for the scheduling process maintained in the maintenance plan. It indicates the future time period (such as one year or 18 months) for which planned dates should be calculated in advance during the scheduling of a maintenance plan. It can be used for both time-based and performance-based maintenance plans.

Shift Factors and Completion Requirement

The shift factor provides the option to advance or delay the next planned maintenance date. If the last planned maintenance work has been completed earlier or with some delay compared to the actual planned date, you can use a shift factor value of 100% to move the next planned date by an equal number of days based on the early or late completion.

If the checkbox for completion requirement is marked in the scheduling parameters section of the maintenance plan, then the next call object (notification or maintenance order or service entry sheet) is only generated after the last call object has been completed.

With the activation of the completion requirement, the subsequent order is generated only after the technical completion of the preceding order (see Figure 5-14). The creation of the next notification is dependent on the completion of the preceding one, in the case of a call object being a notification. The subsequent service entry sheet is generated only after the acceptance of the preceding service entry sheet, in the scenario where the call object is a service entry sheet.

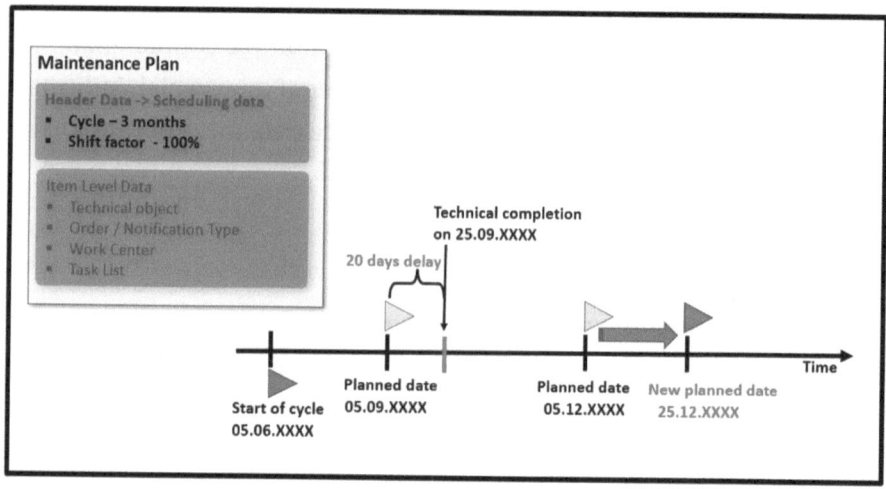

Figure 5-14. *effect of shift factor and completion requirement*

5.3.4. Creating and Scheduling a Single-Cycle Plan with Notification

Similar to generating maintenance orders, a maintenance plan can be set up to create maintenance notifications on a regular basis, based on the frequency recommended by the manufacturer or the organization's maintenance procedure (see Figure 5-15). These notifications are automatically generated from the maintenance plan. For example, sometimes regular maintenance involves only routine inspection of equipment and doesn't require considerable manpower or component changes. Therefore, for such routine inspections, maintenance notifications can be used instead of maintenance orders. If any damage or malfunction is noticed during inspection, the user can create a maintenance order with reference to the notification.

When a maintenance plan is created, the user needs to select the call object as a Notification for the Maintenance Plan category.

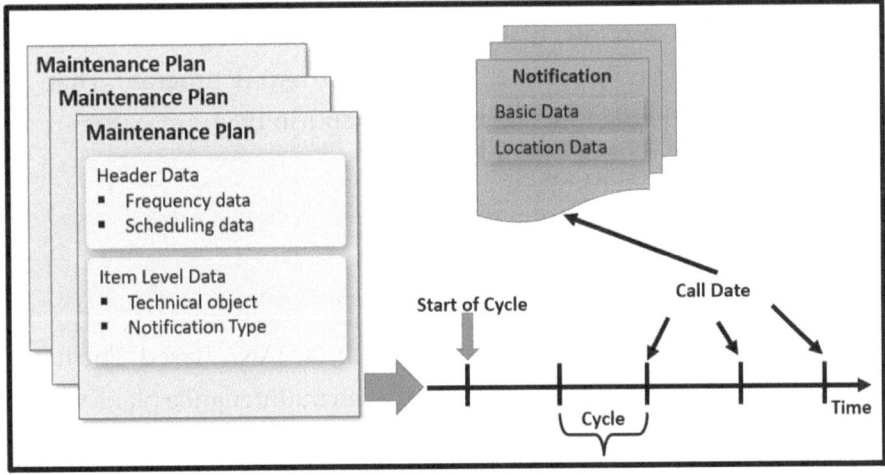

Figure 5-15. *Maintenance planning for notification generation*

5.3.5. SAP GUI-Based (User Interface) Functions for Single-cycle Plans

This section details a few of the important functionalities available in the SAP GUI user interface for a single-cycle plan.

Various transaction codes are available to create maintenance plan in SAP GUI user interface.

- IP41: Create a single-cycle plan

- IP42: Create a strategy plan

- IP43: Create a multiple counter plan

Transaction code IP30 (deadline monitoring) schedules maintenance plans (see Figure 5-16). Multiple maintenance plans can be scheduled together by creating a selection variant and assigning it to each of the maintenance plans. You can define a time period in the Interval for Call Objects field to control the planned waiting dates converted to call objects during execution. Normally, IP30 is executed in the background using a scheduled job that utilizes the IP30 (RISTRA20) ABAP program. The following are scheduling functionalities available in IP30.

- Immediate start for all

- Rescheduling

- Creating call objects

Transaction code IP30H, successor of IP30, is HANA-based. The new transaction code now enables scheduling of all maintenance plans within a specified timeframe at once. When using transaction code IP30H, there is no need to input selection parameter values, as the system internally carries out preselection based on factors like maintenance strategies, dates, and counter readings. This ensures that the system incorporates only the due maintenance plans, eliminating the possibility of missing any call objects. With IP30H, the system selects a significantly smaller number of maintenance plans, leading to improved response times.

The prerequisite to using IP30H is to activate the LOG_EAM_MPSI business function, maintenance plan scheduling, using preselection 1. Reversal of the business function is also possible.

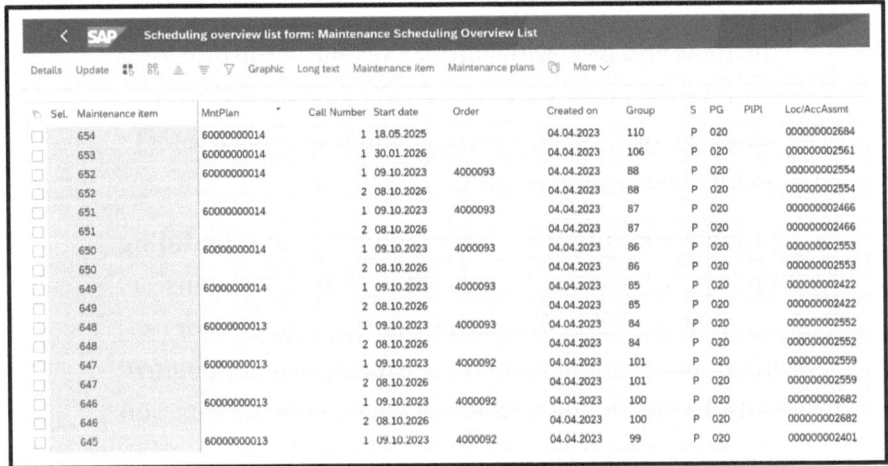

Figure 5-16. *Maintenance plan scheduling overview report*

- **Maintenance plan costing**: You can utilize the cost estimate to calculate the maintenance expenses within a designated timeframe.

- **General costing**: With general costing, you can compute the maintenance expenses for either a single maintenance plan or multiple plans.

- **Object-related cost estimate**: You have the option to derive the maintenance expenses for a technical object over a defined duration. In both maintenance plan general costing scenarios, the cost estimation takes into account active maintenance dates within the specified period, excluding waiting planned dates and skipped call dates. If the specified period extends beyond the last active date, the system simulates the relevant dates.

Work time and material are allocated values based on the rates and prices that are applicable at the time of creating the cost estimate. Varied rates, such as those for future posting periods (e.g., internal service rates), are not taken into account.

- **Change document for maintenance plan scheduling**: Change records can be generated when alterations are made to fields within the maintenance plan or its items, or when adjustments are made to maintenance calls. To enable the creation of change documents, you must activate this feature within the maintenance plan category's customization settings.

- **Work list for notification call object**: When maintenance activities for a technical system are carried out through an order, encompassing system components with varying maintenance cycles, a work list of requests (notifications) can be employed. The distinct maintenance plans generate requests for each corresponding part of the system. At a specified time, all requests pertaining to the technical system are consolidated and transformed into an order. Subsequently, the task lists from the relevant maintenance item are duplicated into the order.

The task's target entity is evident for each operation outlined in the order. Typically, the order's reference object is the higher-level technical system (e.g., the functional location), against which the order expenses are allocated.

5.3.6. Customizing Maintenance Plan

The following describes important customizable objects in a maintenance plan.

- **Maintenance plan category**: This category is used to determine the type of call objects for the maintenance plan.

- **Set pop-up with completion date**: To provide an option to enter two dates while completing a maintenance order: one for entering the completion date and another for entering the reference date.

- **Number ranges**: Number ranges for maintenance items and plans.

- **Sort field**: The sort field is used for grouping multiple maintenance plans so that all plans can be scheduled in one attempt.

- **Default order type for maintenance item**: This allows you to get a default order type during maintenance plan creation.

- **Adjust order type for immediate release**: Provide the option to release the maintenance order immediately at the time of creation from the maintenance plan.

From the SAP Easy Access menu, navigate to Tools → Customizing. Double-click IMG → SPRO–Execute Project. Click the SAP Reference IMG button.

Table 5-5 lists important configuration paths related to plant.

Table 5-5. *Customizing Maintenance Plan*

Configuration Step	Configuration Path
Maintenance plan category.	Plant Maintenance and Customer Service → Maintenance Plans, Work Centers, Task Lists, and PRTs → Maintenance Plans → Set Maintenance Plan Categories
Set the completion date in the pop-up.	Plant Maintenance and Customer Service → Maintenance Plans, Work Centers, Task Lists and PRTs → Maintenance Plans → Set Maintenance Plan Categories.Mark Completion Date checkbox
Assign number ranges for maintenance plans and maintenance items.	Plant Maintenance and Customer Service → Maintenance Plans, Work Centers, Task Lists and PRTs → Maintenance Plans →1. Define Number Ranges for Maintenance Plans2. Define Number Ranges for Maintenance Items
Define the Sort field.	Plant Maintenance and Customer Service → Maintenance Plans, Work Centers, Task Lists and PRTs → Maintenance Plans → Define Sort Fields for Maintenance Plan
Assign a default order type for the maintenance item.	Plant Maintenance and Customer Service → Maintenance and Service Processing → Maintenance and Service Orders → Functions and Settings for Order Types → Define Default Order Types for Maintenance Items
Adjust the order type (Release Immediately).	Plant Maintenance and Customer Service → Maintenance and Service Processing → Maintenance and Service Orders → Functions and Settings for Order Types → Configure Order Types. In the order type definition screen, mark the Release Immediately checkbox.

5.4. Planning Regular External Service Procurement

For the routine maintenance of a technical object carried out by an external vendor, the preventive maintenance aspect of S/4HANA Asset Management provides the ability to create a service entry sheet according to the necessary frequency. These service entry sheets act as notifications to remind about the upcoming regular servicing of a technical object. After the technical object's servicing is completed, the user needs to process the service entry sheet so that the service cost is allocated to the linked maintenance order.

For example, a two-year service package has been procured from the car manufacturer for a car. According to this service contract, the car needs to be sent for servicing every three months. A maintenance plan has been setup and scheduled to generate a service entry sheet every three months throughout the two-year duration.

The following is the process flow for maintenance plan–based service procurement (also see Figure 5-17).

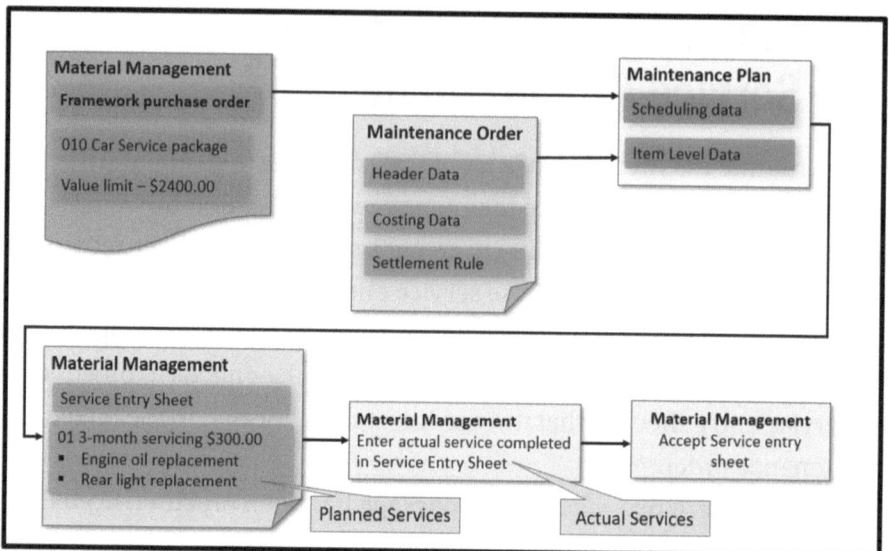

Figure 5-17. *Process flow for maintenance plan-based service procurement*

1. **Create a framework order.** Like any other procurement process, a framework purchase order (category–FO) needs to be created for service procurement in The Materials Management application of S/4HANA. Vendor and value limits need to be maintained. Service specifications are not required.

2. **Create a maintenance order for settlement.** For absorbing the cost of regular service from external vendor, a maintenance order needs to be created which is used for cost collection and cost settlement.

3. **Create a maintenance plan.** Setting up a maintenance plan to have service entry sheets generated regularly based on the necessary frequency. The maintenance plan with category

MM is created with a reference to the framework purchase order and maintenance order for settlement (see Figure 5-18).

A G/L account must be maintained in the plan, which is used for posting the cost of services for financial accounting. In controlling, the cost is updated using the settlement order. Entering service specifications is optional for the performed service. Services from various purchasing documents (such as quotations, purchase requisitions, normal purchase orders, and contracts) can also be entered manually, or a service master record can be maintained.

Pricing is calculated using condition records maintained at various levels: service master record, vendor and service master record, or plant, as well as vendor and service master record.

4. **Scheduling the maintenance plan for service entry sheet generation.** The maintenance plan is scheduled to calculate a planned date when the service is due. When the planned date is reached, a service entry sheet is automatically created for the framework purchase order.

5. **Enter services.** Process the service entry sheet after the completion of regular service. The service entry sheet is updated for the framework order in the event of any discrepancies between the planned and actual services provided, as well as the pricing. The actual values is compared with the value limit specified in the framework order.

6. **Accept the service entry sheet.** After updating the service entry sheet, it is accepted. The cost of the service is updated in both the maintenance settlement order and the G/L account maintained in the maintenance plan.

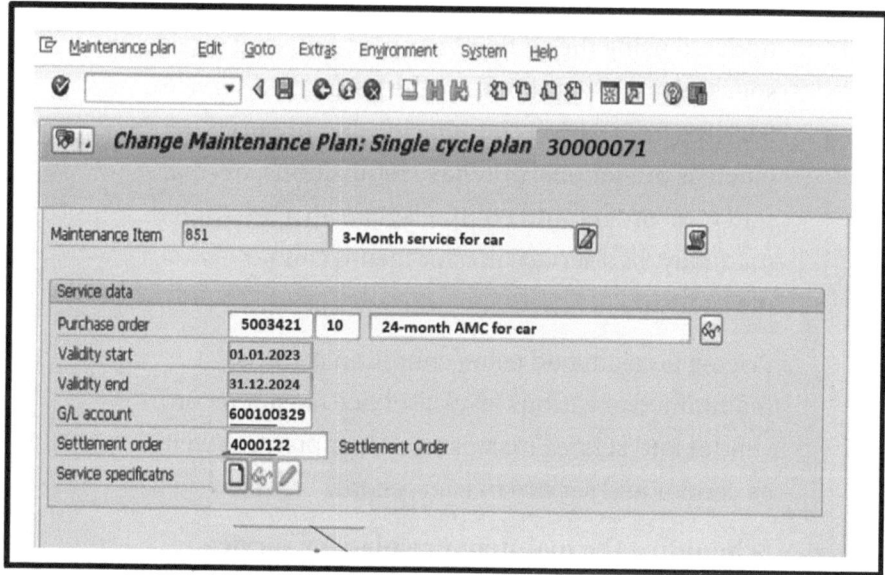

Figure 5-18. *Maintenance plan for service entry sheet generation*

5.5. Maintenance Planning with Time-Based Strategy

Maintenance planning with a time-based strategy deals with planning and generating regular call objects (notifications, maintenance orders) for a technical object consisting of multiple subassemblies and components that require maintenance at different intervals (e.g., monthly or yearly).

A time-based strategy is employed to create and schedule a maintenance plan to generate regular maintenance orders for a technical object with varying frequencies for its different components.

For example, a preventive maintenance plan for an air conditioner may encompass routine cleaning and visual inspections every three months, while tasks such as filter replacement and chemical cleaning occur every six months, with a gas top-up scheduled annually (see Figure 5-19).

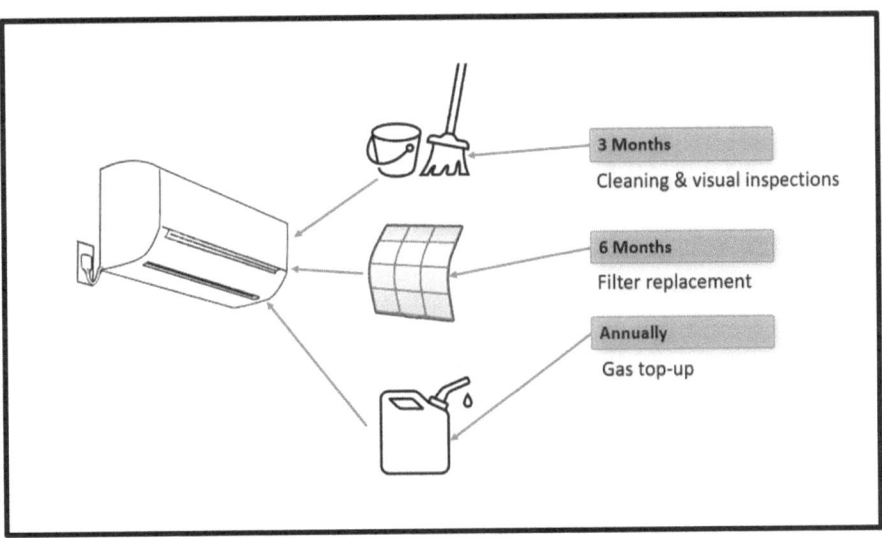

Figure 5-19. *Maintenance planning with time-based strategy*

5.5.1. Maintenance Strategy

A maintenance strategy groups multiple maintenance packages and scheduling parameters for maintenance planning. A maintenance strategy can comprise any number of maintenance packages. All the packages assigned to a maintenance strategy must share the same cycle duration unit (such as months for all packages). For instance, a maintenance strategy consisting of two packages—one with a cycle duration of one month and another with a cycle duration of one year—must be defined as one month and twelve months, respectively.

The following maintenance strategies are assigned to maintenance task lists. The important elements of a time-based maintenance strategy (see Figure 5-20).

- Scheduling indicator

- Call horizon

- Shift factor and tolerances

- Time: factory calendar

- Package sequence and where-used list

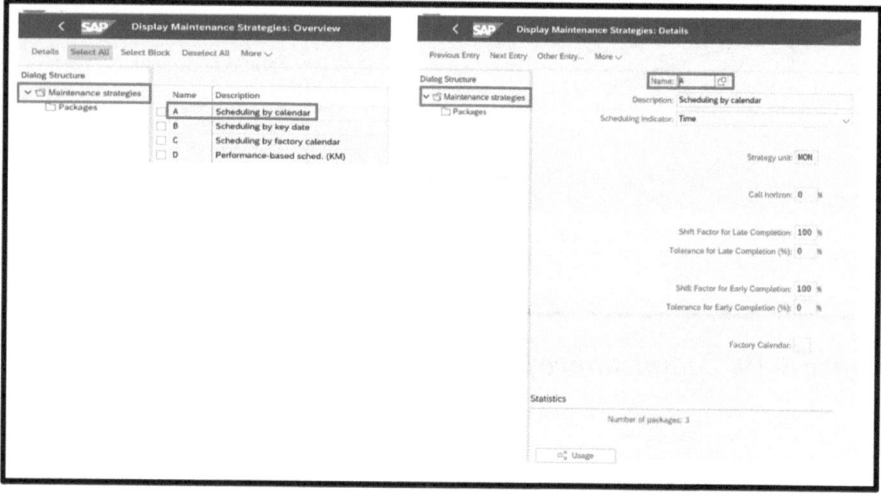

Figure 5-20. *Maintenance strategy and scheduling parameters*

Maintenance Package

Maintenance packages are elements of a maintenance strategy, and they are assigned to the operations of a maintenance task list. Each maintenance package defines a duration frequency at which the operation is to be performed. For example, consider a maintenance task list with two operations: the first operation involves cleaning and visual inspections and is assigned maintenance package-1 (every three months), while the second operation, which includes filter replacement and chemical cleaning, is assigned maintenance package-2 (every six months).

The following lists important parameters for a maintenance package.

- Maintenance package number

- Description

- Cycle length

- Unit of measurement

- Hierarchy

- Offset

- Initial and subsequent buffers

The *offset* of a maintenance package determines the timing for the initial due date of a maintenance package. You must set an offset if the first maintenance should occur at a different time than the regular cycle.

Specified in days relative to the planned date, the *initial and subsequent buffers* establish the start and end dates for a maintenance order.

The *hierarchy* of the maintenance package determines which package to call if multiple packages are due simultaneously. It is assigned to a maintenance package. Depending on the hierarchy, all packages can be called, or some packages can be ignored.

A maintenance package with a higher hierarchy level is selected when multiple packages are due simultaneously. Maintenance packages with the same hierarchy level are called together if all packages are due at the same time (see Figure 5-21).

For example, preventive maintenance for an air conditioner may involve two operations: the first operation involves cleaning the filter, while the second operation includes filter replacement. At six months, both operations are due. To ensure that filter cleaning is not performed before replacement, the second operation is assigned a higher hierarchy level compared to the first operation. As a result, the first operation is not executed when both operations are due.

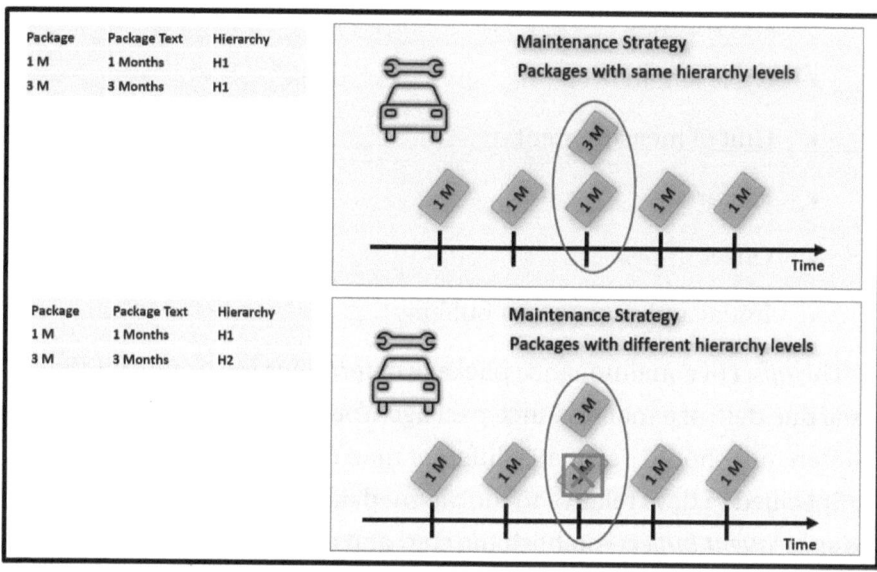

Figure 5-21. *Maintenance packages and the effect of hierarchy levels*

5.5.2. Assigning a Maintenance Strategy to a Task List

In a strategy-based maintenance plan, the user needs to assign the maintenance strategy to a task list that is used to create the strategy-based maintenance plan. The following explains the steps to assign a maintenance strategy to a task list.

During the creation of a task list, the user needs to assign a maintenance strategy to the header section of the task list.

In the next step, the packages in the maintenance strategy are assigned to each operation of the task list (see Figure 5-22). These packages define the frequency of the operations.

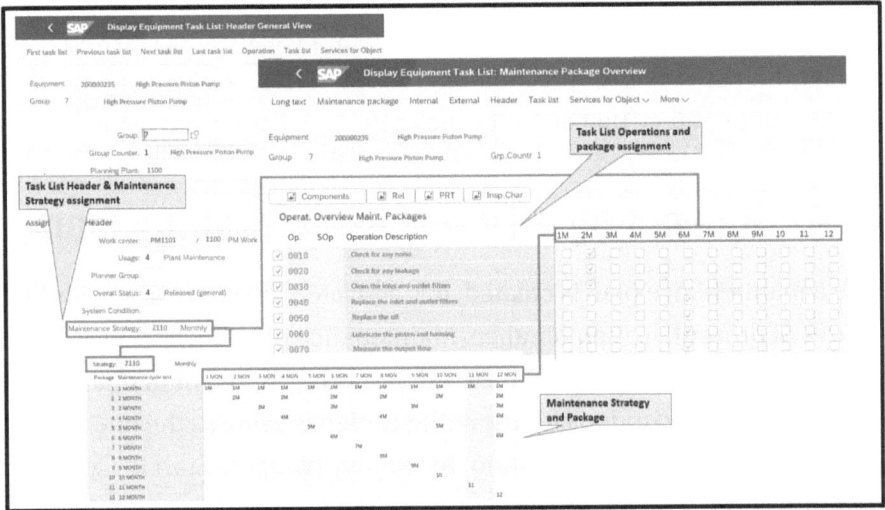

Figure 5-22. *Maintenance package assignment to operations of a task list*

5.5.3. Creating a Time-Based Strategy Plan

A time-based strategy plan follows more than one schedule to generate call objects such as maintenance orders. Maintenance orders generated for different schedules may contain varying operations and components. For example, servicing an air conditioner every three months requires inspection and cleaning, whereas servicing it every twelve months may necessitate the replacement of filters or gas refilling.

A time-based strategy plan is created with a specific maintenance strategy assigned to it. Users can input the task list group and counter directly into the respective fields, or these can be chosen using the search help functionality. In the search help input screen, the maintenance strategy of the plan gets copied automatically as a selection criterion, ensuring that only task lists with the same strategy are selected. The maintenance packages used in the assigned task list are checked and displayed in the cycles for the maintenance plan. The packages of the maintenance strategy that are not used in the task list are not displayed.

5.5.4. Scheduling a Time-Based Strategy Plan

Once a maintenance plan is created, it is activated through the scheduling features of preventive maintenance. Maintenance plan scheduling produces call objects (maintenance orders/notifications) punctually.

The scheduling options for the strategy plan is same as the one for single-cycle plan except one option. In strategy plan, the Start in Cycle scheduling option allows you to start scheduling the strategy plan from the middle of the cycle (see Figure 5-23).

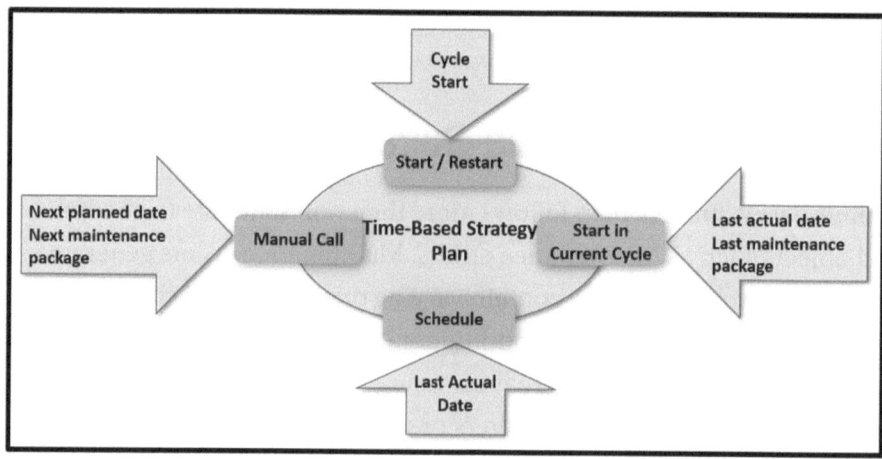

Figure 5-23. Scheduling options in a time-based strategy plan

For example, a maintenance strategy comprises two-month, six-month, and 24-month packages. At the time of maintenance plan creation, preventive maintenance was already in progress in the organization. Four weeks ago, the eighth two-month package was executed. The scheduling option "start in the cycle" enables the maintenance plan to begin not with the first two-month package but with the ninth two-month package due and the third six-month package.

The following describes the scheduling parameters.

- **Factory calendar**: The factory calendar from the maintenance strategy gets copied to the strategy plan as a default. The default factory calendar can be changed in the strategy plan, if required.

- **Cycle start**: Planned and call dates calculation start date.

- **Call horizon**: Call horizon is used to calculate call date, the date when maintenance order, notification or service entry sheet is created before the planned date (actual execution date of the maintenance order, notification or service entry sheet).

- **Scheduling period**: Scheduling period indicates the future time period (such as 1 year or 18 months) for which planned dates should be calculated in advance during the scheduling of a maintenance plan.

- **Completion requirement**: With the activation of the completion requirement, the subsequent order is generated only after the technical completion of the preceding order.

- **Shift factors**: The shift factor provides the option to either advance or delay the next planned maintenance date. If the last planned maintenance work has been completed earlier or with some delay compared to the actual planned date, you can use a shift factor value of 100% to move the next planned date by an equal number of days based on the early or late completion.

- **Tolerance for shift factors**: The tolerance for the shift factor function defines the time duration for which early or delayed completion of a call object compared to the planned date does not affect future plan dates.

- **Cycle modification factor**: One of the scheduling parameters for the maintenance plan is the cycle modification factor. The cycle modification factor allows you to adjust the cycle times for a maintenance strategy on a per-plan basis. If the factor is greater than 1, it extends the strategy's cycle times (see Figure 5-24), while a factor less than 1 reduces them. This factor allows you to lengthen or shorten the maintenance cycle, such as adapting to temporary extra needs while keeping the maintenance strategy unchanged. Importantly, the cycle modification factor is applicable solely to the specific maintenance plan where it has been set.

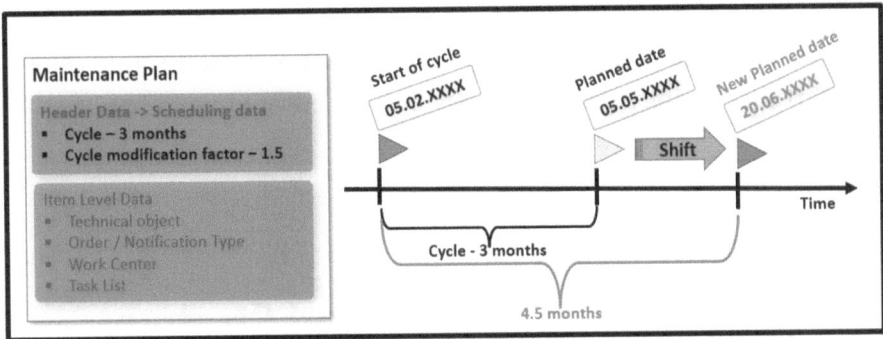

Figure 5-24. *Cycle modification factor*

- **Scheduling indicator**: The scheduling indicator defines the type of time-based scheduling assigned. The following are three types of time-based scheduling.

  - **Time (based on calendar)**: Calendar days are considered for calculation of dates.

  - **Time-Key-Date based**: From the cycle start, the dates are always calculated for the specific key date.

  - **Time (using factory calendar)**: Working days are considered for calculation of dates.

5.5.5. SAP GUI-Based (User Interface) Functions for Strategy Plans

Engineering change management functionality can be used for maintenance and service task lists. Different task list versions can be saved with various change masters. Change masters prove to be particularly advantageous when specific combinations of maintenance packages remain valid for only a limited duration. In such instances, the change master maintains a task list version with a specific validity date. When a task list associated with a change master is utilized in a maintenance plan, the validity date of the change master is cross-referenced during plan scheduling. If a maintenance package no longer holds validity (or is not yet valid) concerning a given plan date, it is deactivated for that specific plan date.

5.6. Maintenance Planning with Performance-Based Strategy

Maintenance planning with a performance-based strategy deals with the planning and generation of regular call objects (notifications, maintenance orders) for a technical object consisting of multiple subassemblies and components that require maintenance at their different running performance (such as 1000 hours, 25000 hours). For example, a preventive maintenance plan for diesel generator may encompass routine cleaning and visual inspections every 1000 hours and tasks such as filter replacement and oil replacement occur after every 25000 hours (see Figure 5-25).

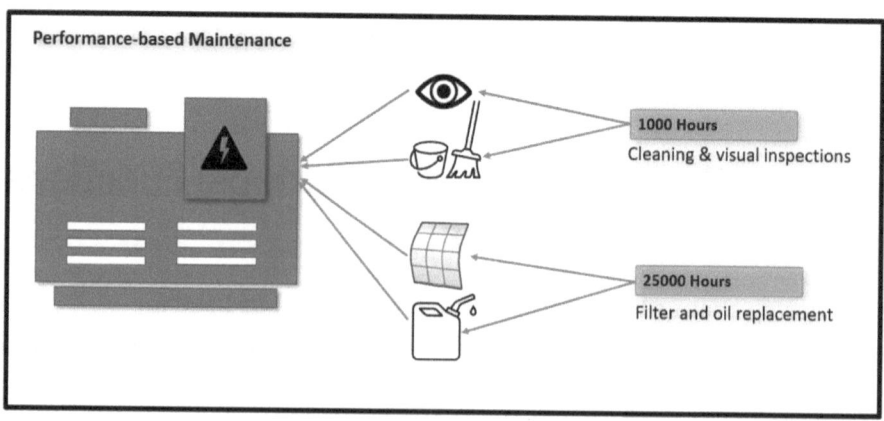

Figure 5-25. *Maintenance planning with a performance-based strategy*

Preventive maintenance can be scheduled and executed through a performance-based approach, utilizing regularly input counter readings. The calculated maintenance dates are automatically adapted according to the entered counter readings. Master data element measurement counter is required to create for the technical object (equipment, functional location master).

5.6.1. Counters and Measurement Documents

Master data counters are used to consistently record incremental values associated with the operation of a technical object. They are always established with reference to a technical object. For instance, a counter is utilized to log the total operating hours of equipment on a daily basis.

A counter is linked to a characteristic (such as operating hours, flow, or volume) from the classification system. This characteristic is always associated with the corresponding characteristic unit (e.g., hours or liters). To create a counter, you must maintain a fixed estimated annual aggregated reading value. This value serves as a reference for calculating intervals for maintenance plan dates.

The first measurement document entered for a newly added counter is known as the initial measurement document. It indicates the current counter reading at that particular moment. In the absence of an initial measurement document for the new counter, a maintenance plan that relies on this counter cannot be started. Performance-based maintenance planning relies on measurement documents. To reflect the maintenance scheduling process with maximum accuracy, input the measurement documents consistently (see Figure 5-26) and as frequently as feasible.

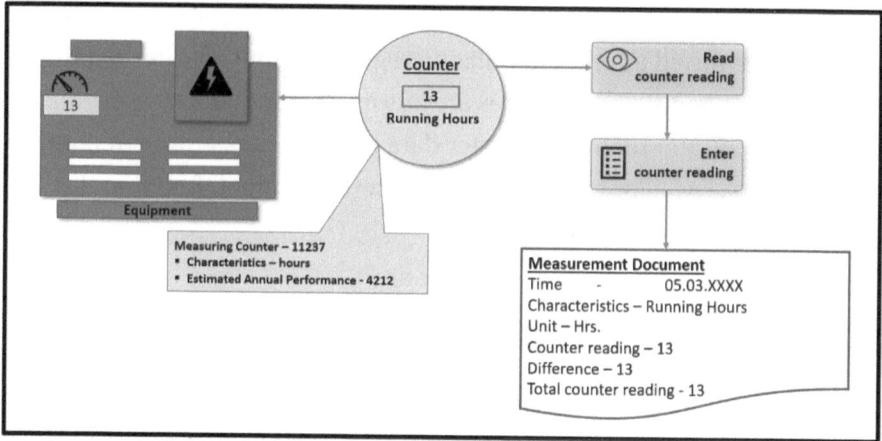

Figure 5-26. *Measuring counter and creating initial measurement document*

A performance-based maintenance strategy is created using the Performance Scheduling Indicator (activity). The performance value unit is assigned in the strategy. A maintenance strategy can comprise any number of maintenance packages. The packages are comprising of same parameters as in time-based strategy and packages such as maintenance package number, cycle length, unit of measurement and others.

5.6.2. Creating a Performance-Based Maintenance Strategy Plan

A performance-based strategy plan follows more than one schedule to generate call objects such as maintenance orders. Maintenance orders generated for different schedules may contain varying operations and components. For example, servicing a diesel generator every 1000 hours of running requires inspection and cleaning, whereas servicing it every 25000 hours of running may necessitate the replacement of filters and oil.

Performance-based strategy plans are of two types: the single-cycle plan and the strategy plan.

A measuring counter is required in the technical object for maintenance planning using a performance-based maintenance plan. Like a time-based maintenance plan, the maintenance plan category and maintenance strategy need to be assigned in a performance-based maintenance plan. Upon entering the reference object, the counter is automatically suggested, determined by the unit aligned with the maintenance strategy (e.g., hours). In the case of a performance-based single-cycle plan, the counter is automatically suggested once the maintenance cycle and reference object are provided.

Users can input the task list group and counter directly into the respective fields, or these can be chosen using the search help functionality. In the search help input screen, the maintenance strategy of the plan gets copied automatically as a selection criterion, ensuring that only task lists with the same strategy are selected. The maintenance packages used in the assigned task list are checked and displayed in the cycles for the maintenance plan (see Figure 5-27). The packages of the maintenance strategy that are not used in the task list are not displayed.

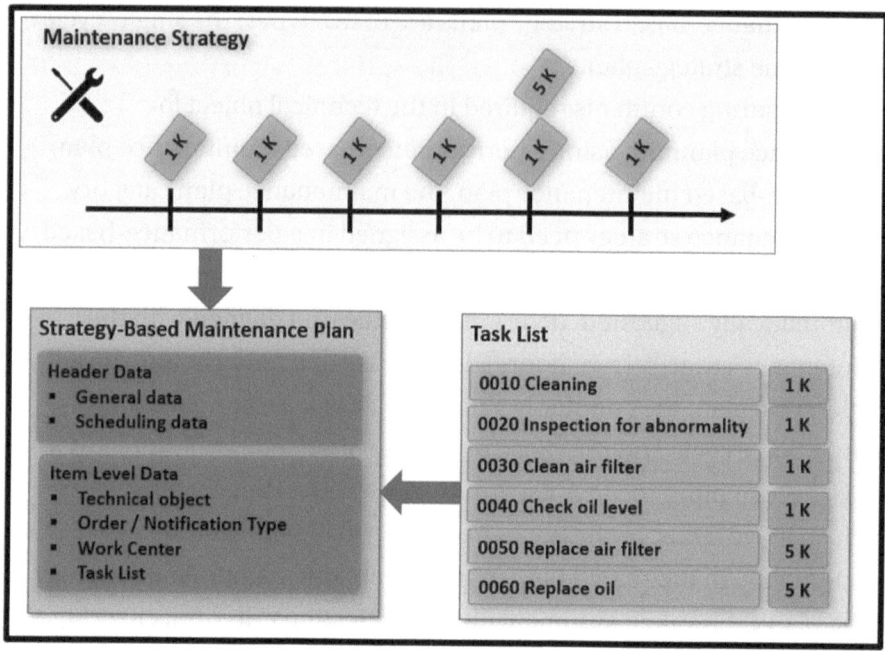

Figure 5-27. *Creating a performance-based maintenance strategy plan*

5.6.3. Scheduling a Performance-Based Maintenance Strategy Plan

The scheduling options for the performance-based strategy plan are the same as the ones for the single-cycle plan except for one option. In the performance-based strategy plan, the Start in Cycle scheduling option allows you to start scheduling the strategy plan from the middle of the cycle (see Figure 5-28).

Entering a counter reading greater than the one in the latest measurement document is not permissible, as maintenance planning relies on the current data recorded in the system through measurement documents.

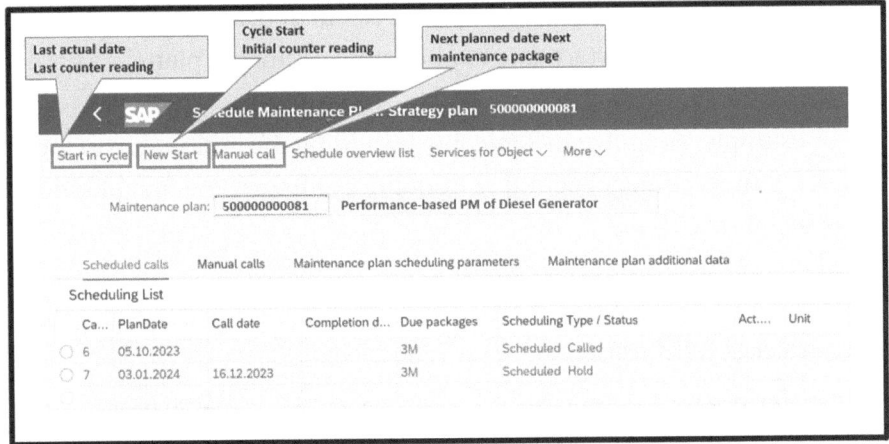

Figure 5-28. *Scheduling options in a performance-based strategy plan*

The following describes the scheduling parameters.

- Call horizon

- Scheduling period

- Initial counter reading

- Completion requirement

- Shift factors

- Tolerance for shift factors

- Cycle modification factor

Basic Scheduling Functions

Performance-based maintenance plans are scheduled according to the projected yearly performance input in the reference technical object counter (see Figure 5-29). Based on the projected annual performance, the daily performance is calculated. Using the calculated daily performance and maintenance cycle, the time period is calculated and then added to the maintenance plan's start date to obtain the estimated plan date based on the current daily performance.

For example, as shown in Table 5-6, by adding the cycle value to the current counter reading, beginning from the cycle start, the first planned date is set after 50 days.

Table 5-6. *Counter Reading and Cycle Value*

Projected annual performance	7300 hours
Daily performance based on annual performance	7300 hours/365 days = 20 hours per day
Planned date for a cycle of 1000 hours	1000/20 = 50 days

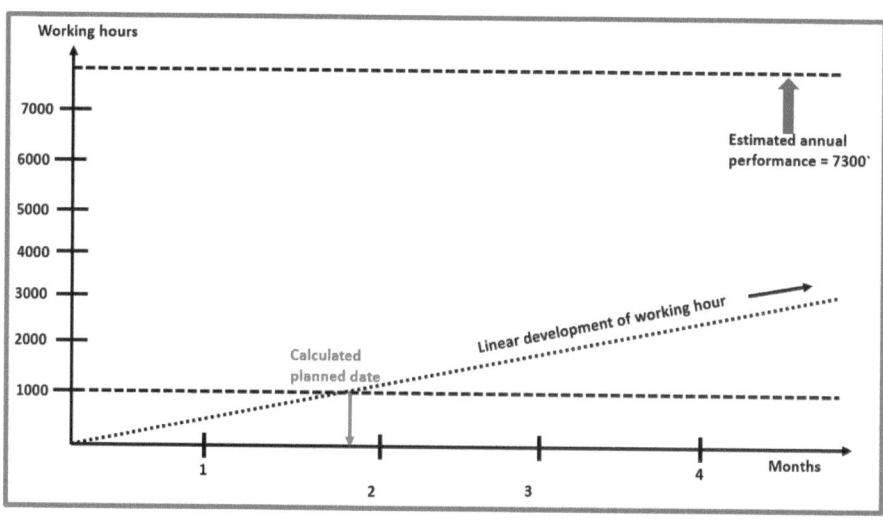

Figure 5-29. *Projected planned date based on estimated annual performance*

A plan date is calculated depending on the cycle and projected annual performance. If a measurement document is created, it leads to the recalculation of the planned date. In simpler terms, the planned dates are influenced directly by the entered measurement documents (see Figure 5-30). This implies that the regular input of measurement documents is necessary to achieve a planned date that accurately mirrors the performance value.

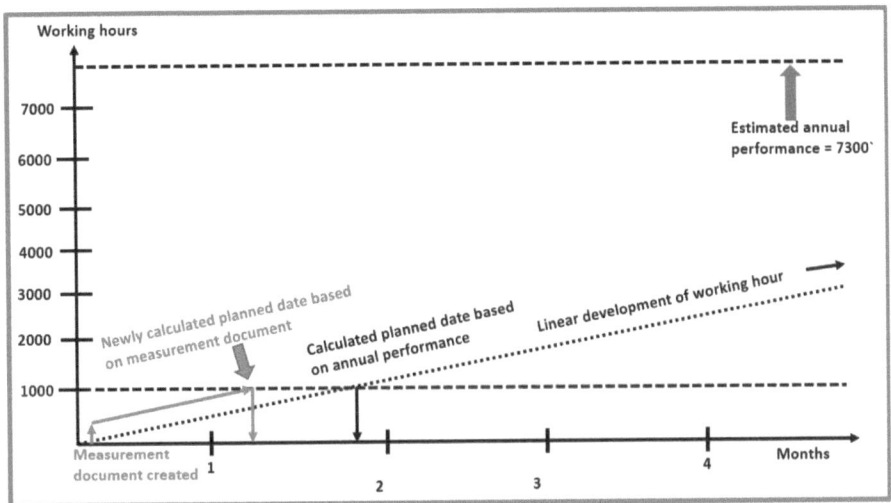

Figure 5-30. *Planned date adjustment based on regular entry of counter reading*

Call Horizon

A call horizon calculates call date, the date when maintenance order, notification or service entry sheet is created before the planned date (actual execution date of the maintenance order, notification or service entry sheet). It is specified in percentage and refers to the duration of the maintenance cycle.

Let's look at the following data as an example.

- The estimated annual performance of a diesel generator is 7300 hours.

- The daily performance based on annual performance is 20 hours (7300 hours/365 days = 20 hours).

- The preventive maintenance is 1000 hours of running.

- The planned date for a cycle of 1000 hours is 50 days (1000/20 = 50)

- The maintenance plan's cycle starts at a counter reading of 0 hours on 05.03.XXXX.

- The project's planned date for maintenance is 24.04.XXXX

- The call horizon entered is 90%

- The call date for creating the call object (maintenance order) is 19.04.XXXX (90% of 50 days = 45 days; therefore, 50 – 45 days = 5 days before the plan date).

For illustrative purposes, let's assume that regularly created measurement documents contain an average of 20 hours per day. When running hours vary daily, the planned date and the call date are recalculated accordingly.

If the call horizon is left unspecified during the scheduling of the maintenance plan, the system takes it to be 0%. Consequently, a call is initiated right away (meaning an order is generated) as soon as the maintenance plan kicks off, regardless of the counter reading. If you intend to maintain the cycle duration, it's recommended to establish a call horizon of 100%.

5.7. Maintenance Planning with Cycles of Different Dimensions

A multiple counter maintenance plan consists of functionalities to plan and schedule regular maintenance for a technical object that needs to be maintained based on cycles of different dimensions, such as total running hours and output generated in liters (see Figure 5-31). For example, in a filtration plant, the water pump's regular maintenance is scheduled for every 2160 hours of running or 500,000 liters of suction.

To plan and schedule regular maintenance for water pumps in a clarification plant, a multiple counter plan can be used, where maintenance orders is generated based on cycles of different dimensions, such as after 2160 hours of running or 500,000 liters of suction.

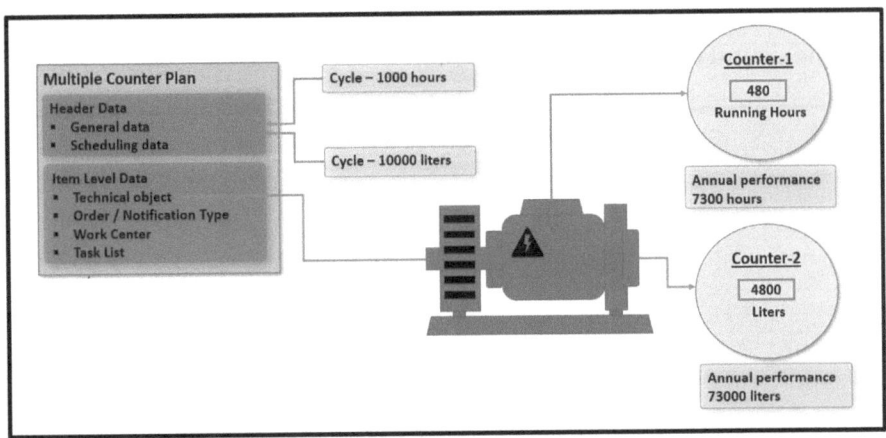

Figure 5-31. *Multiple counter plans*

5.7.1. Creating and Scheduling a Multiple Counter Plan

In a multiple counter plan, cycles are used to maintain the required performance value. The maintenance strategy is not employed in this plan. Cycles can be assigned freely to a multiple counter plan. If a strategy is assigned to the entered task list in a multiple counter plan, the strategy holds no relevance during the operation selection from the task list.

The starting of the multiple counter plan relies on the present counter readings, with the date of the measurement document containing the latest counter value being the determining factor. The scheduling is consistently adjusted according to the current counter readings.

Restarting a multiple counter plan is possible. The "Start in current cycle" function is appliable only in multiple counter plans featuring sequences of cycle sets. The manual call option does not applies to the multiple counter plan (see Figure 5-32).

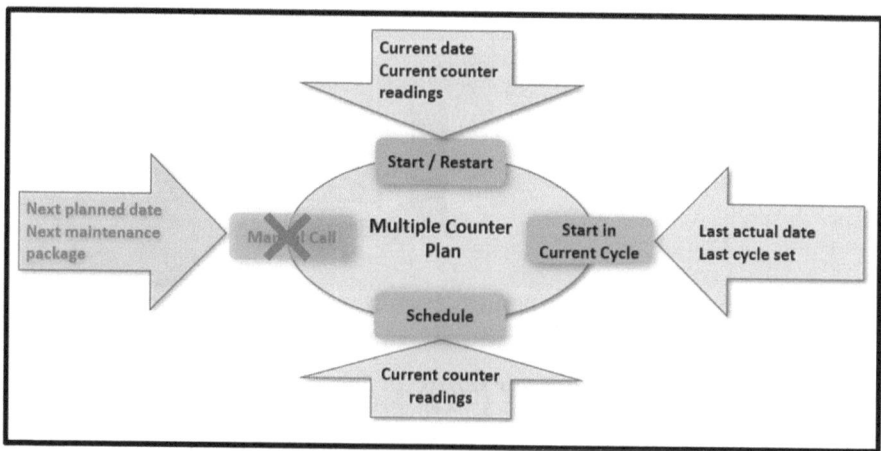

Figure 5-32. *Scheduling function in multiple counter plan*

The following describes the scheduling parameters of a multiple counter plan.

- **Cycle modification factors**: The cycle modification factor allows you to adjust the cycle times for a maintenance plan. If the cycle modification factor is greater than 1, it extends the cycle times. If it's less than 1, it reduces them. This factor enables you to either lengthen or shorten the maintenance cycle.

- **Preliminary buffer**: In a multiple counter plan, you can input a preliminary buffer in terms of the number of days. This buffer indicates the lead time before the planned maintenance date when the order should be generated.

- **Operation type**: Operation type defines the link type for maintenance cycles. Choosing an AND link results in the order being generated only when all cycles reach the required performance value. On the other hand, with an OR link, the order is generated as soon as any of the cycles reach the required performance value.

- **Start date and time**: Through deadline monitoring, a maintenance plan started based on the designated start date and time.

- **Scheduling period**: The scheduling period indicates the future time period (such as 1 year or 18 months) for which planned dates should be calculated in advance during the scheduling of a maintenance plan.

- **Completion requirement**: By activating the completion requirement, the subsequent order is generated only after the technical completion of the preceding order.

- **Shift factors**: The shift factor offers the choice to either advance or postpone the next planned maintenance date. If the last scheduled maintenance task has been carried out before or after the actual planned date, you can employ a shift factor value of 100% to adjust the next planned date by the same number of days, contingent on the early or delayed completion.

- **Tolerance for shift factors**: The shift factor tolerance function establishes the time duration during which the early or delayed completion of a call object compared to the planned date does not impact the future planned dates.

The multiple counter plan is triggered based on the current counter reading. The initial projected plan date is determined using the reading from the relevant measurement document (see Figure 5-33). In the case of an AND link type, the projected plan date is calculated based on the largest interval originating from the cycle and performance per day.

In a multiple counter plan, to achieve precise scheduling in accordance with the cycle sets, the following list of process flow steps should be followed.

1. Create the maintenance order.

2. Complete the maintenance order.

3. Record the measurement document upon order completion or consistently enter measurement documents at regular intervals.

4. Reschedule the multiple counter plan.

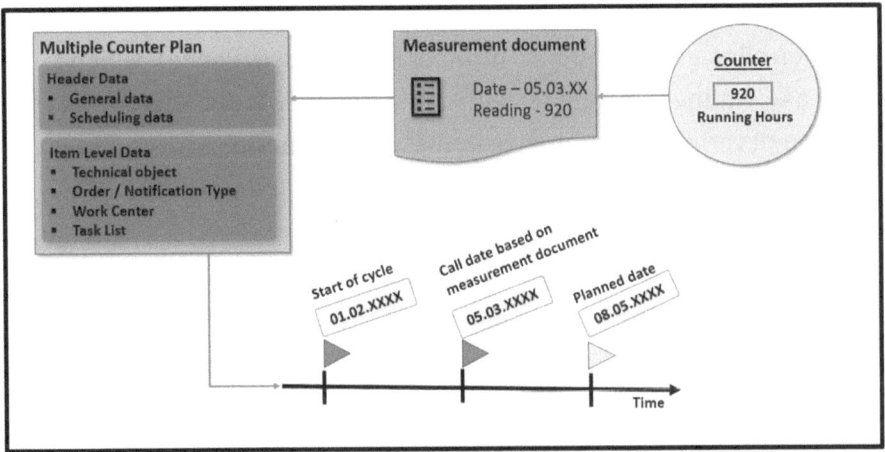

Figure 5-33. *Posting measurement document*

Unlike the performance-based strategic plan, the scheduling of the multiple counter plan doesn't involve high numbers of cycle counts for establishing the new planned date. Instead, scheduling is consistently computed based on the current counter readings.

The actual call date is ascertained from the entered measurement documents, which increase in the counter reading. If the maintenance cycle becomes due as a result of these measurement documents, the call is promptly initiated. When the cycle isn't due as per the measurement document, the call is initiated on the calculated planned date.

5.7.2. Using Cycle Set Sequence

The cycle set is used as a template to generate a multi-counter plan. Cycle set comprises of successive maintenance cycles (see Figure 5-34). Unlike the maintenance strategy, the cycle set lacks a reference function. Consequently, once the maintenance plan is created, there is no longer a connection with the cycle set. When the cycle set becomes part of the maintenance plan, changes or removals can be made to individual cycles.

Also, multiple cycle sets can be used in a multi-counter plan. Each cycle set is then allocated a sequence number (cycle set sequence), which aids in associating the cycle set with a maintenance item. This allows the inclusion of various maintenance items with distinct cycles. It's important to note that only one time-based cycle can be utilized within a multi-counter plan.

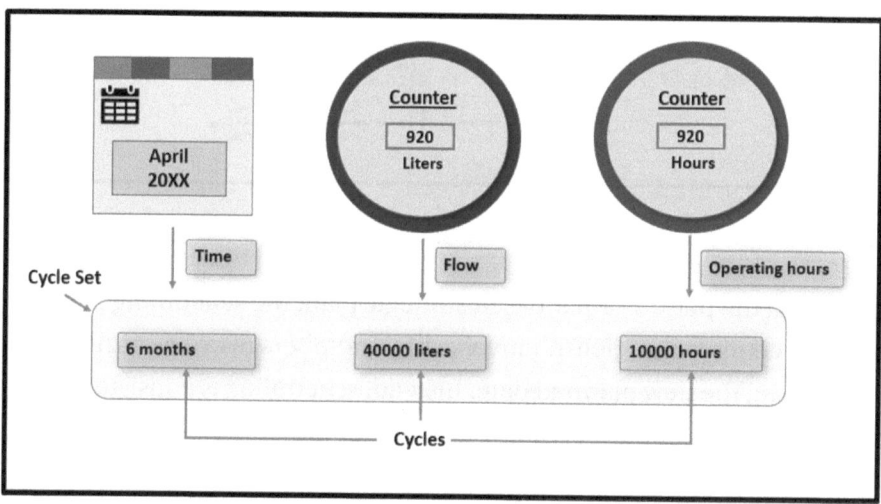

Figure 5-34. *Cycle sets and cycles*

For example, a diesel generator set is scheduled for regular maintenance after 1800 hours of running and/or three months. Regular maintenance includes tasks such as cleaning and visual inspection. After another 1800 hours of running and/or three months, a detailed maintenance will be carried out. The detailed maintenance includes tasks such as replacing the air filter and oil.

In a multiple counter plan, using two cycle sets with two cycles in each cycle set, you can plan regular maintenance for the diesel generator in the example. Each of the cycle sets needs to be assigned to an item in the plan, one for regular maintenance and the other for detailed maintenance.

Common multiple counter plans typically operate with individual cycles connected by an operation type (OR or AND). These cycles usually involve the same maintenance item, which means the set of operations is always the same. However, multiple counter plans are not designed to accommodate different maintenance strategies. Thus, unlike the functionality of maintenance packages within a maintenance strategy, they cannot execute a different set of operations.

To address this constraint, *cycle set sequences* were developed. These sequences combine multiple cycles into a set that can be allocated to a maintenance item. The concept revolves around integrating multiple cycle set sequences within a single multiple counter plan, enabling a scheduling pattern akin to 1-2-1-2. By assigning specific maintenance items to either "1" or "2," it becomes possible to generate varying tasks.

The following describes the scheduling process.

1. When calculating the first plan date, scheduling takes into account cycle set 1 and generates a maintenance order (call object) only for this date, corresponding to the maintenance item linked with cycle set 1.

2. Using this date as a reference, scheduling evaluates the cycles within cycle set 2 to calculate the subsequent plan date. Once again, it generates a call object only for the pertinent maintenance item.

3. The scheduling process revisits the first cycle set after scheduling the second and, consequently, the highest cycle set.

5.8. User Interface

S/4HANA Asset Management comes with several Fiori apps to help user perform preventive maintenance planning and execution. The following are important Fiori apps.

- Fiori Apps–Process Task List (Planner), Fiori App ID–W0021: Three apps are available for Create Task List, Change Task List, and Display Task List.

- Fiori Apps–Find Maintenance Task List, Fiori App ID–F2660: This app lets you find and display the Equipment task list, Functional location task list, and General maintenance task list.

- Fiori Apps–Manage Maintenance Items, Fiori App ID–F5356: This app helps a maintenance planner manage maintenance items, including create maintenance items, assign maintenance item to a maintenance plan, view details of a maintenance item.

- Fiori Apps–Find Maintenance Items, Fiori App ID–F3621: This app lets you find and display maintenance items.

- Fiori Apps–Find Maintenance Plans, Fiori App ID–F3622: This app lets you find and view maintenance plans, its assigned maintenance items, its properties, and its scheduled and manual maintenance calls.

- Fiori Apps–Manage Maintenance Plans, Fiori App ID–F5325: With this app, you can manage maintenance plans.

- Fiori Apps–Mass Schedule Maintenance Plans, Fiori App ID–F2774: As a maintenance planner, you can use this app to schedule all maintenance plans due within a specific time frame.

5.9. Summary

This chapter explored various aspects of maintenance planning and execution.

Master data includes maintenance task lists and their applications, user interfaces, and essential customizations. The single-cycle maintenance plan, creation, and scheduling of single-cycle maintenance plans, along with insights into customizing these plans to suit specific needs. The planning of regular external service procurement focuses on generating a service entry sheet through a maintenance plan, streamlining the process of external service procurement.

Maintenance planning with a time-based strategy explains the concept of a maintenance strategy and guides crafting and scheduling time-based strategy plans. Furthermore, maintenance planning with a performance-based strategy and the creation and scheduling of performance-based maintenance strategy plans were outlined, offering insights into optimizing maintenance activities.

The chapter delved into maintenance planning with cycles of different dimensions, detailing the intricacies of developing and scheduling multiple counter plans accommodating various maintenance cycle dimensions. Finally, the user interface was explored, highlighting SAP Fiori apps designed for working with task lists and maintenance plans.

CHAPTER 6

Costing and Budgeting

This chapter is about money and how it functions when attending to assets like equipment, machines, and manufacturing assembly lines. Imagine you have an instrument that's broken and requires repair. Fixing things incurs costs, and this chapter helps you understand where the money goes when technical assets are being fixed.

The following are some of the key topics covered.

- Costing in asset maintenance and repair

- Maintenance order cost settlement and closure

- Budgeting in asset maintenance and repair

6.1. Costing in Asset Maintenance and Repair Process

In S/4HANA Asset Management, financial costing in maintenance orders is a way to track and manage the expenses related to maintenance activities in a business. It helps organizations keep a close eye on the costs incurred during maintenance, such as servicing equipment, repairing machinery, or any other maintenance-related activities.

© Rajesh Ojha and Chandan Mohan Jaiswal 2023
R. Ojha and C. M. Jaiswal, *SAP S/4HANA Asset Management*,
https://doi.org/10.1007/978-1-4842-9870-1_6

6.1.1. Integration with Finance and Controlling

As part of logistics, asset management is closely integrated with other business areas, such as finance (accounting). This implies that the organizational units defined for finance and controlling in S/4HANA also cover aspects of costing in asset management. For example, the creation of asset management–specific cost centers with annual plans.

The following lists the most common organizational objects and their structure in S/4HANA Finance and Controlling (also see Figure 6-1).

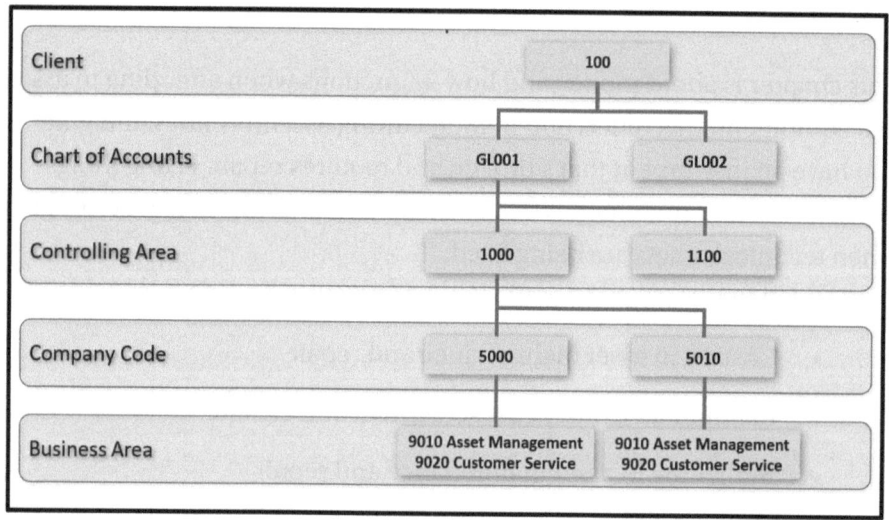

Figure 6-1. *Organizational objects and their structure in Finance and Controlling*

- **Client**: The client stands as the top-level entity among all organizational units. For example, it may represent a conglomerate of corporations comprising multiple companies. All the business applications within a client share a common database for data.

- **Controlling area**: a controlling area in S/4HANA is a way of grouping and organizing one or more company codes to enable centralized control and management accounting functions across multiple subsidiaries or business units within an organization.

- **Chart of accounts**: In the SAP S/4HANA system, each company is associated with a specific chart of accounts, and all G/L (General Ledger) accounts used by that company are mapped to this chart of accounts. The G/L accounts within a chart of accounts are distinct and not repeated.

- **Company codes**: A company represents an independent legal entity or a business unit within a corporate group. For instance, if you have a parent company with multiple subsidiaries, each subsidiary may be represented as a separate company code in S/4HANA.

- **Cost centers**: A cost center is like an account/bucket where all the expenses related to a specific function, department, and project are collected and monitored (see Figure 6-2). It helps organizations understand how much money is being spent in different parts of the business.

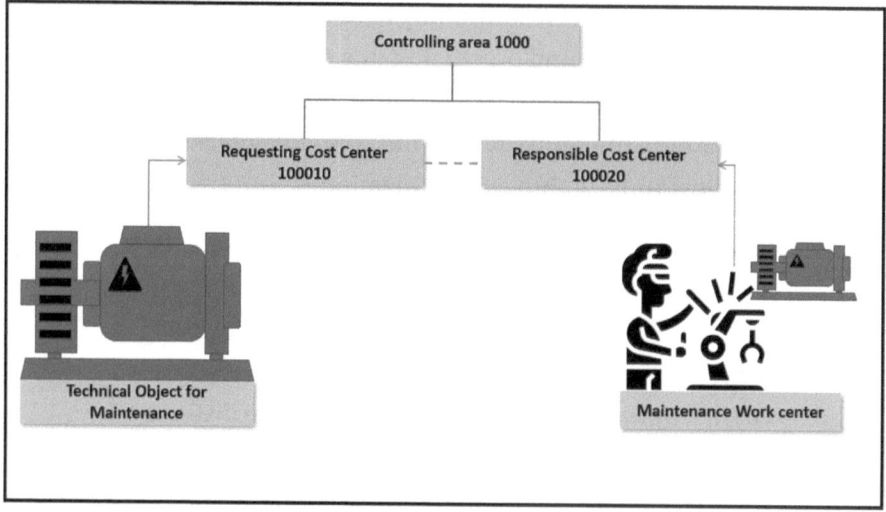

Figure 6-2. *Cost centers in S/4HANA Asset Management*

Cost centers are created within a controlling area. For asset management, a minimum of two cost centers are relevant to capture and monitor internal costs (see Figure 6-3).

- A **responsible cost center** is assigned to the maintenance work center, which performs maintenance and repair activities for an asset. It is assigned in the Costing view of the maintenance work center.

- A **requesting cost center** is created to collect and monitor the costs incurred on the asset. This cost center is entered in the Organizational view of the equipment master or functional location of an asset.

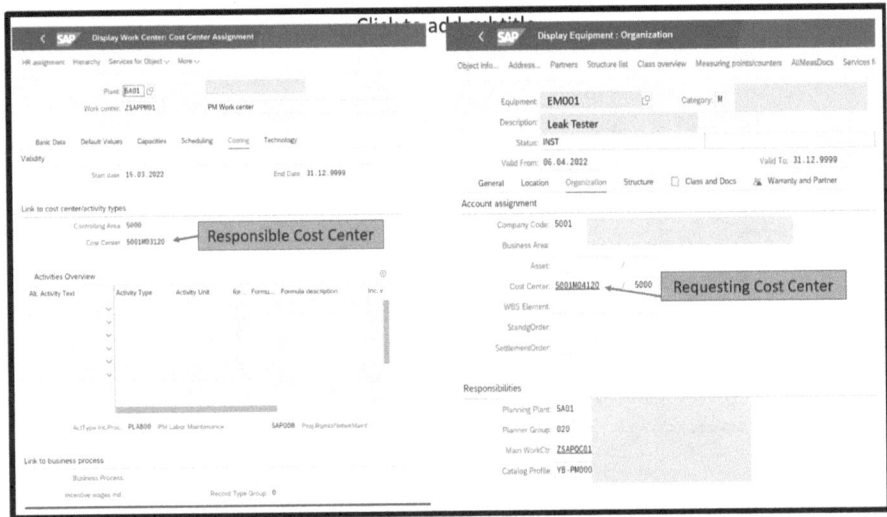

Figure 6-3. *Responsible and requesting cost centers*

6.1.2. Various Accounting Terms (Order Values) in Maintenance Order

Maintenance and repair work in a maintenance order generates and incurs various financial and cost values, which can be classified according to costs incurred, output (activities) generated, and settlements. These financial and costing values (see Figure 6-4) also depend on the maintenance order type and processing phase.

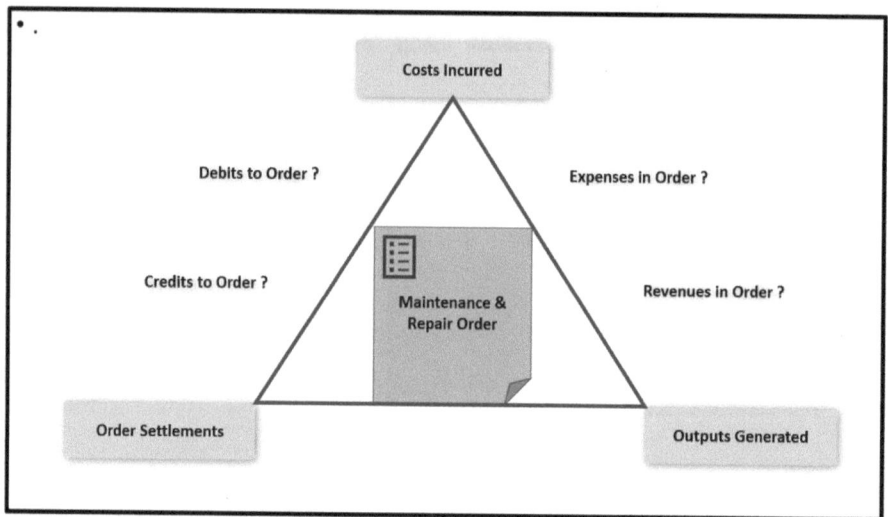

Figure 6-4. *Various account terms (order values) in maintenance order*

- **Costs** are incurred when spare parts consumed, third-party services rendered, and internal man-hours utilized during the maintenance and repair work of an asset, and these get posted to the maintenance order. As a result, the maintenance order gets debited with the value of the cost posted.

- **Activities** are the output generated during an asset's maintenance and repair work. For example, refurbishment of an assembly of a technical system adds value to the assembly by improving its working condition, which, in turn, results in credits being posted to the refurbishment maintenance order, and the order gets credited.

- **Settlement** is a process of cost allocation, fully or partially, from one cost object to another. By settling the maintenance and repair orders, you transfer the costs from the order to the relevant cost centers or other cost objects where these expenses belong. This way, the financial reports become accurate, and you can understand the true financial impact of the maintenance activities related to asset management.

- **Expenses** are the costs incurred in a maintenance and repair order and generally represent expenses in the books of accounts. On the other hand, the activities or output generated during maintenance and repair work generally do not represent revenue because they are not sold to customers.

6.1.3. Costing Value Flow in Maintenance Order: Estimated, Planned and Actual Cost

In repair and maintenance orders, there are three types of costs available.

- **Estimated cost**: The value for the estimated cost is entered manually in the repair and maintenance order. Users can enter this value based on their own estimation of spares and man-hours required during the maintenance activity. The estimated cost must be entered before releasing the order. Estimated costs are also saved in the Plant Maintenance Information System (PMIS) for use in reporting and analytics.

- **Planned cost**: The maintenance order-related resources, such as operations with internal man-hours and materials, generate planned costs for the order when these resources are entered in the order and the order is released (dispatched). The planned cost is derived automatically based on the cost estimate, and it is not possible to manually maintain the planned costs.

- **Actual cost**: The maintenance order is automatically debited with actual costs due to resource consumption, such as man-hour utilization confirmation, materials issue posting, and so on (see Figure 6-5). It is not possible to enter actual costs manually.

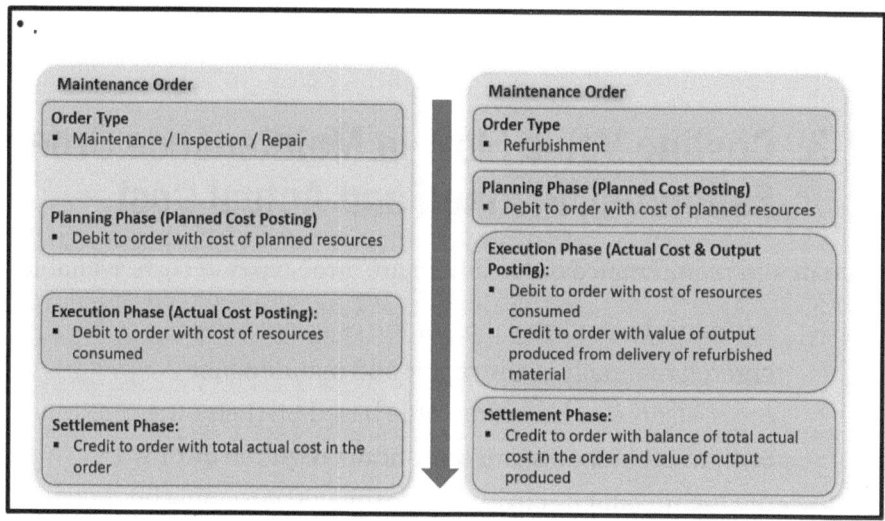

Figure 6-5. *Costing value flow in maintenance and repair order*

The maintenance order types also determine the types of accounting value flow.

During maintenance, inspection, and repair of a technical object using related order types, the order gets debited with planned costs when the required resources, such as spare parts and man-hours, have been planned and the order has been released (dispatched). The order is debited again with the actual cost when the planned spare part has been consumed, or the utilized man-hours have been confirmed. Finally, at the time of order settlement during the completion phase, the debited cost of the order is transferred to the received cost object, such as the cost center.

In refurbishing any assembly or technical object using a refurbishment order type, the order gets debited with planned costs when the required resources, such as spare parts and man-hours, have been planned, and the order has been released (dispatched). The order is debited again with the actual cost when the planned spare part has been consumed, or the utilized man-hours have been confirmed. At the time of delivery (transferring the refurbished technical object to inventory), the order is credited with the equivalent value of the newly refurbished material (technical object). Finally, at the time of order settlement during the completion phase, the difference between the order's debited cost and the output generated (value of the newly refurbished material) is transferred either to the material (settlement receiver for the refurbishment order type) or the price different account.

6.1.4. Option of Costing at Operation Level of Maintenance Order

In general, all the costing values (such as planned and actual costs of spare parts and man-hours) within a maintenance order are derived and stored at the maintenance order level (header level) and not at the individual operation (task) level. However, with the functionality of an operation account assignment (OAA), it is possible to derive all the costing values at each operation level of a maintenance order, enabling the option and flexibility to plan and control costs at the operation level. To use the OAA functionality, the business function, LOG_EAM_OLC, needs to be activated.

Regulators and legal requirement bodies prefer to capture asset management costs at a detailed level so that they can be reported in categories other than the company's overall business strategy. Various industries using asset management software also prefer to perform costing at the individual operation level and derive the aggregated value at the entire job (order) level. With OAA functionality, these needs can be fulfilled. It allows planning and capturing all costing values at each maintenance order operation.

Note that an order can be either header or operation-level, but not both. To distinguish a maintenance order with OAA, the system status activity account assignment (ACAS) is set, which can be viewed in the order or any order-related report. The costing data (such as planned and actual costs of spare parts and man-hours) can be viewed for an operation in a separate view (see Figure 6-6). All costs are only recorded at the operation level and dynamically aggregated at the maintenance order level.

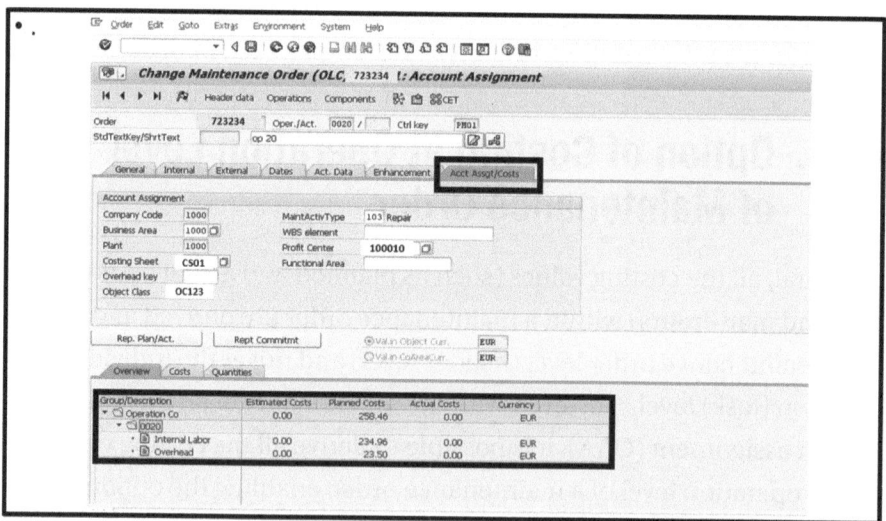

Figure 6-6. *Costing at the operation level of maintenance order*

OAA has various functionalities in the asset management process.

- **Estimated cost**: The user can manually enter the estimated cost only at the operation level, and it is aggregated at the maintenance order level. To facilitate the user, the functionality to copy planned cost to estimated cost can be used by pressing the Copy Estimated Cost button before the maintenance order release.

- **Spare parts issues**: Planning and issuing stocked spare parts make the operation an account assignment for debiting the cost of the spare part. For non-stocked spare parts, the operations are entered as account assignments during goods receipt for the purchase order.

- **Procurement**: When a purchase requisition is generated for a non-stocked spare part required for an operation, the operation gets assigned as the account assignment object in the purchase requisition for the spare part. Subsequently, the OAA is transferred to the purchase order created for the purchase requisition.

- **Settlement**: The settlement rule is maintained at the operation level and not at the order level, and settlement happens for each operation individually.

6.1.5. Overhead Rates

Overhead is a lump sum calculated based on the incurred direct cost. Typically, the calculation involves a percentage of the base cost or an amount-based markup. It is added to the planned and actual cost of a maintenance order to factor in overhead costs consisting of other resources used in maintenance and repair, such as consumable items (lubricant and fasteners), disposable gloves used by technicians, or any other items consumed that are difficult to quantify and plan (see Figure 6-7).

In addition to allocating overhead rates, they distribute sales and administration overhead costs. Furthermore, these overheads can be applied to account for the expenses of resources consumed in the order that are not itemized separately, including administration, transportation, and more.

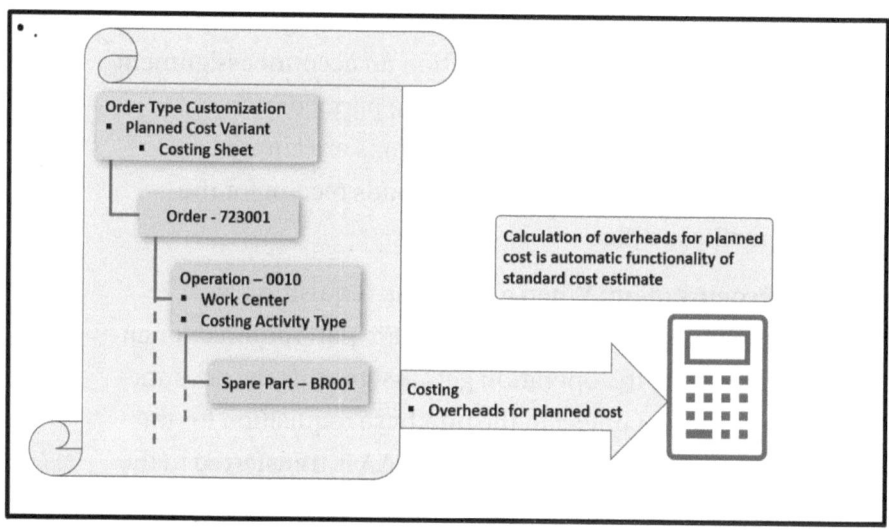

Figure 6-7. *Overheads based on planned cost*

For the total planned cost of maintenance and repair orders, overhead is calculated as part of the standard cost estimate. In other words, it is automatically calculated when the order is saved after planning resources. Additionally, it can be recalculated as and when required, such as when there is a change in the planned quantity of resources. The prerequisite configuration of assigning the costing sheet to the respective planned costing variant for the maintenance order type must be set up in the finance and controlling area to get overhead on planned costs.

Costing Sheet and Costing Variant

The costing sheet and variant are configurable data in finance and controlling applications. The costing sheet links all the parts of the overhead calculation. It comprises baselines, overhead lines, and total lines processed during overhead calculation. The costing variant contains control parameters for costing, determining how costing is carried out. It forms the link between the application and customizing, as all cost estimates are created and saved with reference to a costing variant.

Unlike overheads for planned cost, the overheads for maintenance order actual cost are not calculated automatically. For each consumption of resources (such as spare parts and man-hours), you must run overhead calculations manually. Then the order gets debited with overheads on actual cost (see Figure 6-8). Similar to overheads for planned cost, to obtain overheads on actual cost, the prerequisite configuration of assigning the costing sheet to the respective actual costing variant for the maintenance order type must be set up in the finance and controlling area.

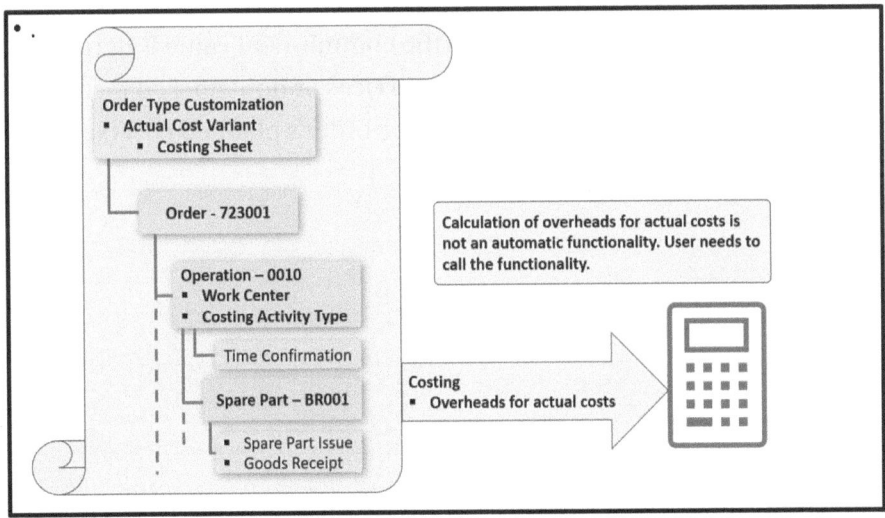

Figure 6-8. *Overheads based on actual cost*

6.1.6. Commitment Management

Commitments in SAP maintenance orders are obligations for external purchases or material requirements planning. These commitment values are not directly recorded in accounting objects, but they can result in actual costs in the maintenance order due to various business transactions.

Commitment management is employed to input these obligations during a maintenance order's early planning stage and examine how they impact costing and finance, particularly concerning purchase orders.

Maintenance order commitments separately update and track specific resources, especially those needed for external procurement. These commitment values are updated in parallel with maintenance order planning and execution. In simpler terms, both the planned and actual costs influence the commitment value, allowing better control and visibility over the resources involved in the maintenance process.

You must activate open items management for the required order type to set up commitment functionality for maintenance orders.

The commitment value gets calculated when the planned cost for the maintenance order is determined. As the commitment value is derived for external resources, such as external services or non-stocked materials, it is typically smaller than the maintenance order's planned cost (see Figure 6-9).

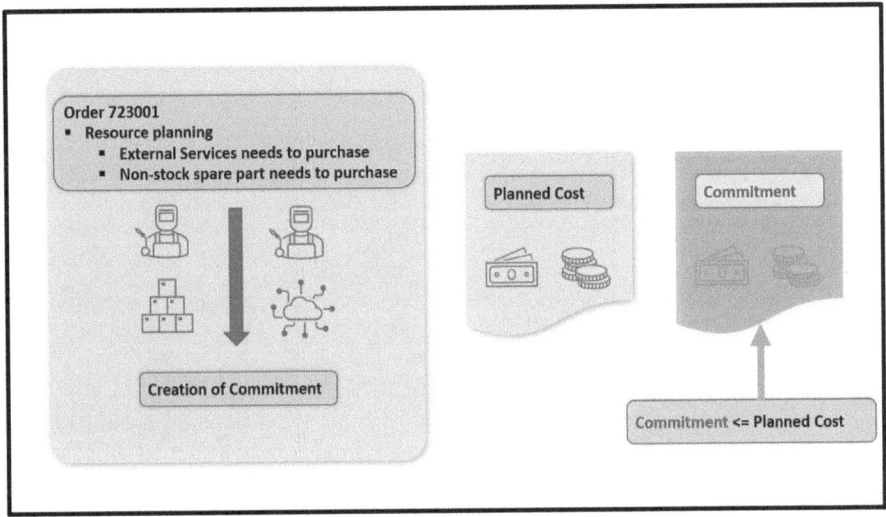

Figure 6-9. *Generating commitment*

However, the commitment value can increase due to certain business transactions after the maintenance order has been planned. These transactions may include generating a purchase order for a purchase requisition linked to an operation in the maintenance order, increasing the price in the purchase order, manually creating a purchase order, and using the maintenance order as a cost object (account assignment) in the purchase order. These business transactions increase the commitment value, which may exceed the planned cost (see Figure 6-10).

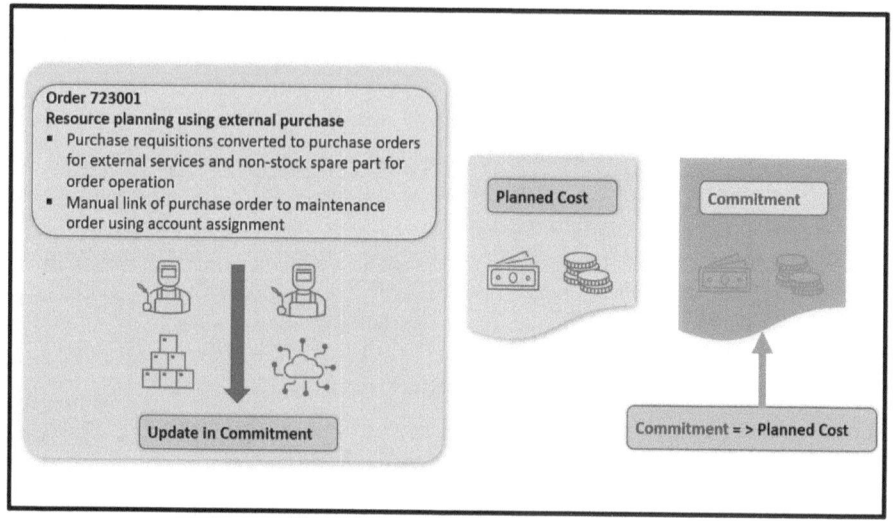

Figure 6-10. *Update in commitment value*

During the processing of the maintenance order, as external resources are consumed—for example, posting goods receipts against a purchase order for non-stock material—the order's commitment value is reduced. The actual costing values determine the reduction in commitment value. Logically, the commitment value should reach zero when all the planned external resources have been fully consumed.

6.1.7. Costing in Refurbishment of Subassemblies

In refurbishing subassemblies/technical objects using refurbishment order type, the costing values flow is different from the maintenance and repair work using maintenance order types.

In the refurbishment order type, along with estimated, planned, and actual cost values, the value for output also gets updated and displayed separately. Here, output refers to the refurbished/overhauled subassembly. The value for output generated is displayed as a negative figure in the refurbishment order and delivery costing reports.

At the time of delivery (transferring the refurbished technical object to inventory), the order is credited with the equivalent value of the newly refurbished material (technical object).

Also, at the time of order settlement during the completion phase, the difference between the order's debited cost and the output generated (value of the newly refurbished material) is transferred to the material (settlement receiver for the refurbishment order type).

6.1.8. Required Configuration Activities for Costing in Maintenance and Repair Orders

Configuration is the process of customizing and setting up various functionalities and features of the SAP S/4HANA software according to the specific needs and requirements of an organization's business processes.

Customizations (configurations) related to costs in maintenance and repair activities in asset management are grouped within many application areas. Such as Organizational structure customization is within the finance and controlling area. The following are important customizations for maintenance order costing.

- Value categories

- Assign cost elements to value categories

- Version for cost estimates

- Default values for value categories

- Costing sheet

- Valuation variant

- Costing variant

- Commitment management

- Settings for costs at an operational level

From the SAP Easy Access menu, navigate to Tools → Customizing. Double-click IMG → SPRO–Execute Project. Click the SAP Reference IMG button (see Figure 6-11).

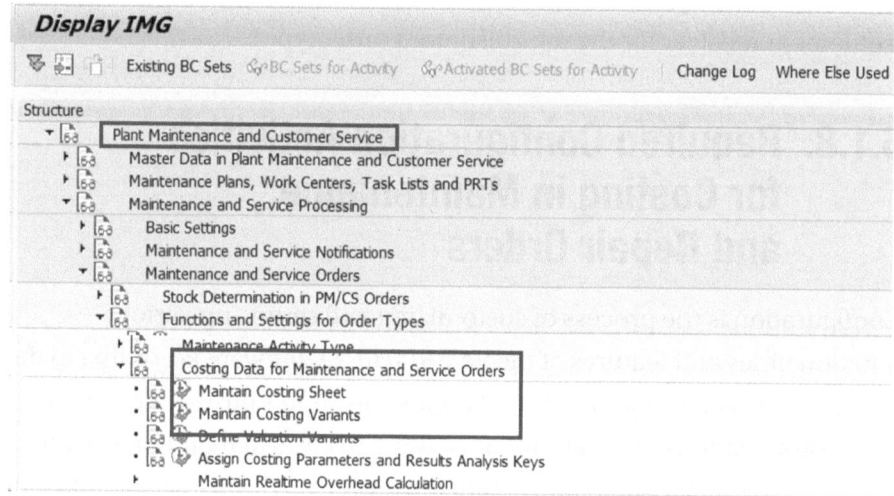

Figure 6-11. *Customization (configuration) nodes for maintenance order costing*

Table 6-1 lists important configuration paths related to maintenance order costing.

Table 6-1. *Configuration Path for Important Configuration Related to Maintenance Order Costing*

Configuration Step	Configuration Path
Maintain value categories.	Plant Maintenance and Customer Service → Maintenance and Service Processing → Basic Settings → Settings for Display of Costs → Maintain Value Categories
Assign cost elements to value categories.	Plant Maintenance and Customer Service → Maintenance and Service Processing → Basic Settings → Settings for Display of Costs → Assign Cost Elements to Value Categories
Assign version for cost estimates.	Plant Maintenance and Customer Service → Maintenance and Service Processing → Basic Settings → Settings for Display of Costs → Define Version for Cost Estimates for Orders
Assign default values for value categories.	Plant Maintenance and Customer Service → Maintenance and Service Processing → Basic Settings → Settings for Display of Costs → Define Default Values for Value Categories
Define the costing sheet.	Plant Maintenance and Customer Service → Maintenance and Service Processing → Maintenance and Service Orders → Functions and Settings for Order Types → Costing Data for Maintenance and Service Orders → Maintain Costing Sheet
Define/assign costing variants.	Plant Maintenance and Customer Service → Maintenance and Service Processing → Maintenance and Service Orders → Functions and Settings for Order Types → Costing Data for Maintenance and Service Orders → Maintain Costing Variants

(continued)

289

Table 6-1. (*continued*)

Configuration Step	Configuration Path
Define the valuation variant.	Plant Maintenance and Customer Service → Maintenance and Service Processing → Maintenance and Service Orders → Functions and Settings for Order Types → Costing Data for Maintenance and Service Orders → Define Valuation Variants
Activate commitment management (OpenItem M.).	Plant Maintenance and Customer Service → Maintenance and Service Processing → Maintenance and Service Orders → Functions and Settings for Order Types → Configure Order Types
Define the settings for costs at the operational level.	Plant Maintenance and Customer Service → Maintenance and Service Processing → Maintenance and Service Orders → Functions and Settings for Order Types → Costs at Operation Level

6.2. Maintenance Order Cost Settlement and Closure

In asset management, the maintenance and repair order is an object for planning, controlling, and executing maintenance activities. All costs incurred from planning and consuming resources (such as materials, man-hours, and external services) are posted to the maintenance order. The order gets debited with the actual costs of the resources consumed.

During the completion phase of the maintenance order, the order is settled to an actual cost receiver, which means the order is credited by allocating its debit to the receiving cost object (see Figure 6-12).

For example, in a manufacturing industry, the settlement receiver can be the cost center of the technical object which was repaired or the cost center of the organizational unit (such as the cost center of the maintenance plant), which requested for a maintenance job. This means the maintenance order collects all the resource costs for reporting and analysis at the maintenance order level. Subsequently, the cost is transferred to receiving cost objects where these expenses belong. This way, the financial reports become accurate, and you can understand the true financial impact of the maintenance activities related to asset management.

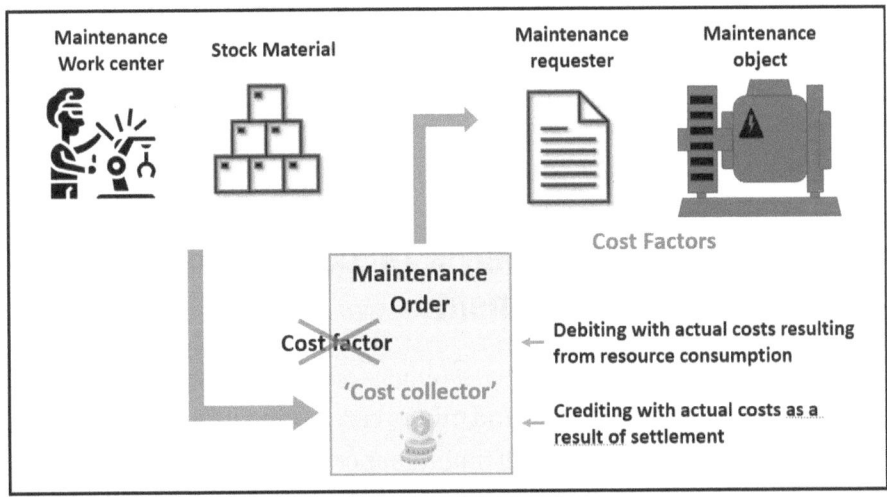

Figure 6-12.   *Settlement process*

6.2.1. Maintenance Order Settlement and Business Completion

Figure 6-13 illustrates settlement for continuous tasks (in-progress orders).

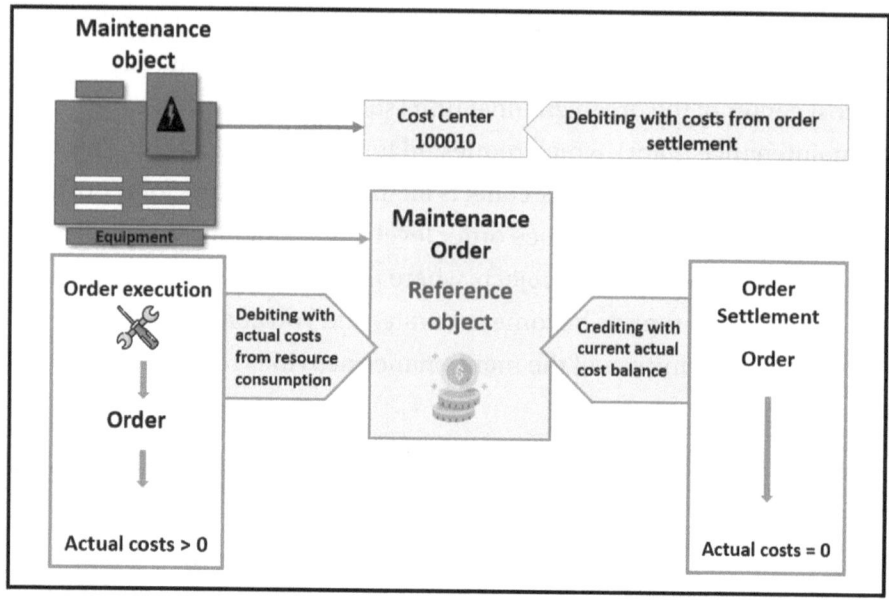

Figure 6-13. *Settlement for continuous tasks (in-progress orders)*

Types of Settlement Receiver (Account Assignment Objects) in Order Settlement

Apart from the cost center, other cost objects such as the work breakdown structure (WBS), assets, orders, and others can also be used as settlement receivers in the settlement rule of maintenance and repair orders. The nature of the maintenance and repair activity typically determines the type of settlement receiver (see Figure 6-14).

For example, maintenance and repair activities for shutdown maintenance can be settled to WBS or network. Maintenance activities related to improvement/investment projects for an asset, which increases the asset's value, should be settled to the asset. On the other hand, repair activity costs that need to be claimed from any vendor/supplier of the asset can be settled in an internal order. In customer service, where you repair customers' assets and later raise an invoice to receive the price of the repair, such maintenance and repair orders should be settled to the sales order.

Based on the configuration for the settlement rule in a maintenance order type, the appropriate settlement receiver type is entered automatically in the maintenance and repair order.

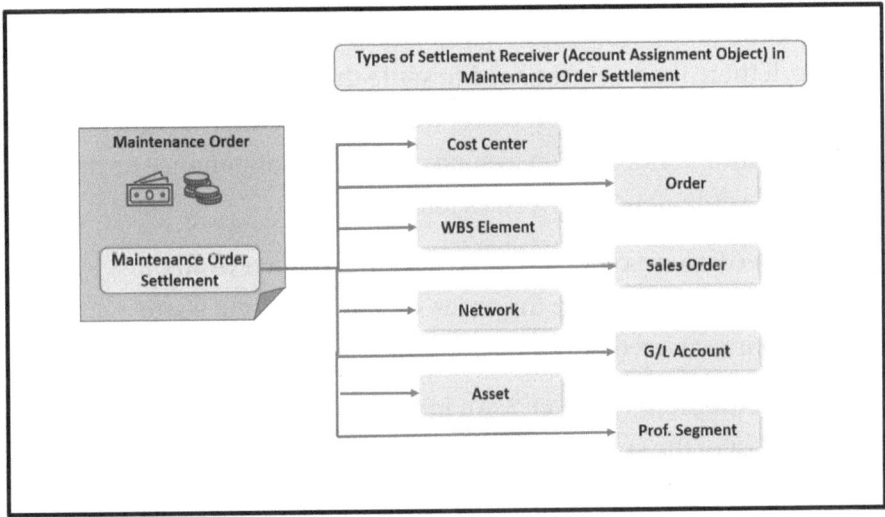

Figure 6-14. *Various types of settlement receivers (account assignment objects) in an order settlement*

Options to Derive Settlement Rule in Maintenance Order

The settlement rule in a maintenance order can be entered in various ways, such as the following (also see Figure 6-15).

- Manually, by calling the settlement rule maintenance screen in the maintenance order

- Automatically, when the maintenance order is released or technically completed

- Manually with or without system default (In a system default, the settlement receiver is copied automatically from the technical asset's equipment master data.)

The time the settlement rule is entered depends on the configuration setting for settlement rule timing. But at the latest, the settlement rule must be entered by maintenance order technical completion time. In general, one settlement rule is entered in a maintenance order, but if required, multiple settlement rules can also be entered in a maintenance order. For example, if there is a business need to settle the cost of a specific spare part to a different settlement receiver (such as in an internal order created specifically for a vendor), then in such cases, the maintenance order is set up with two settlement receivers, as follows.

- **Settlement receiver 1**: A cost center to receive all the debited costs from the order, except the cost of spare parts covered under warranty.

- **Settlement receiver 2**: An internal order to receive the cost of the spare parts covered under warranty.

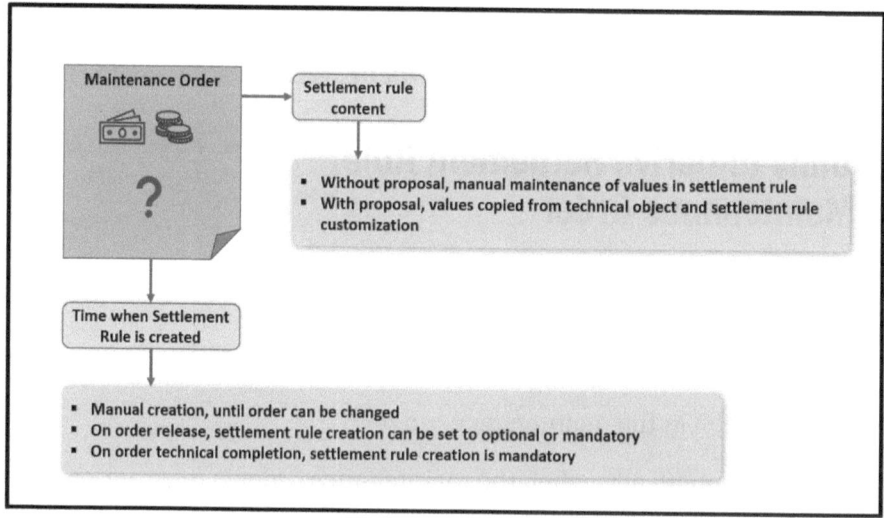

Figure 6-15. *Options to derive settlement rule in maintenance order*

An order can be settled multiple times if required during maintenance and repair. For example, for a large maintenance activity spanning over multiple months, the maintenance order is settled at the end of every month if it has a value of debited cost more than zero during each settlement. The settlement is part of the period-end activity performed by finance and controlling. Therefore, during each period-end settlement, if a maintenance order is available with cost, then the order is settled to transfer the total cost to a receiver cost object and reduce the cost to zero in the maintenance order (see Figure 6-16).

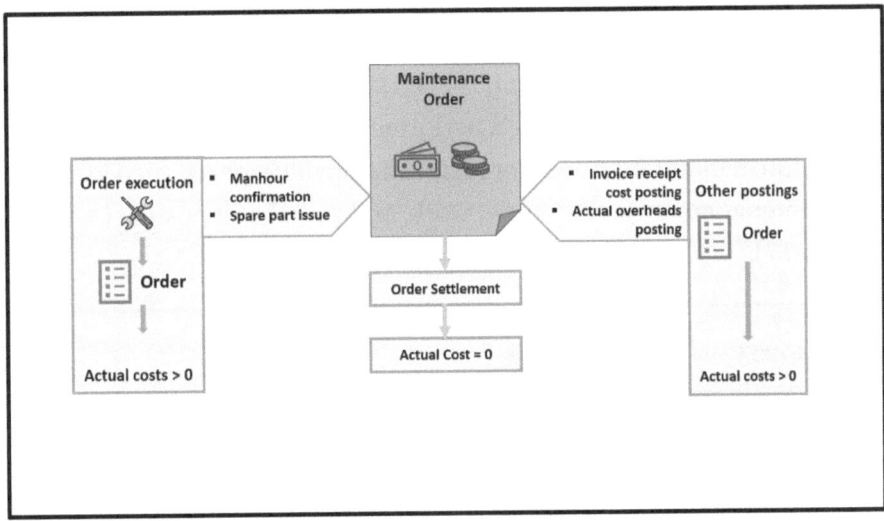

Figure 6-16. *Order settlement*

Order Completion Phase

The maintenance and repair order completion phase consists of several steps. First, the order is completed from a technical point of view (technically completed) in the asset management application area after all the maintenance operations have been finished. If a settlement rule is not defined in the order by that time, then during technical completion, the settlement rule is entered in the maintenance order. Once the maintenance order is technically completed, no further changes are possible.

An order technically completed means no further maintenance and repair is expected. As part of period-end activities, settlement can be performed for the order to transfer the actual debited cost to the receiving cost object. Still, the already settled maintenance order can accept costs, such as for the external vendor's incoming invoice posting for the external service purchased during the repair work in the order. In this case, the maintenance order needs to settle again during the future period-end settlement process.

After the maintenance order has been settled and there is zero actual cost debit, business completion is performed for the order. After business completion, a maintenance order is closed from all aspects, such as repair and maintenance work, settlement work, and actual cost posting (see Figure 6-17).

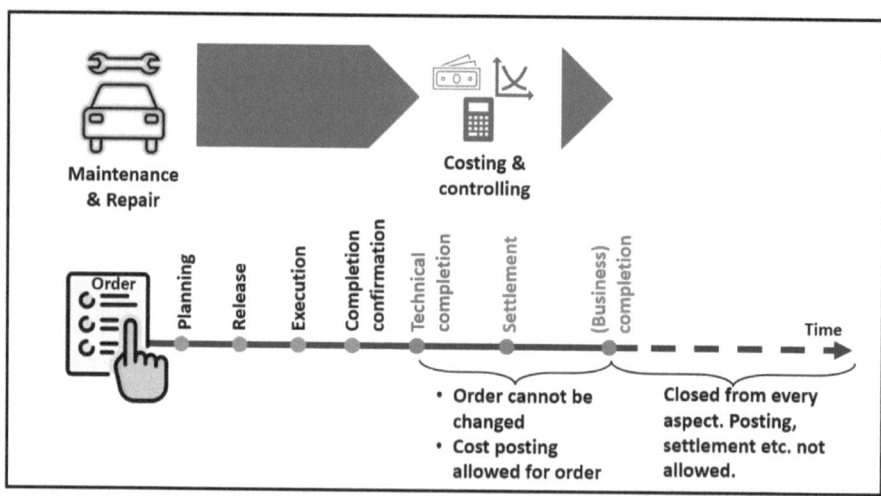

Figure 6-17. *Allowed business function after completion of a maintenance order*

The Document Flow functionality in the maintenance order displays a hierarchical structure, listing the documents created and their status during maintenance order execution. These documents can be opened from the document flow, and preceding and succeeding documents related to the opened documents can also be accessed. For maintenance and repair orders, the following is the list of documents that appear in the document flow.

- Maintenance notification

- Maintenance plan

- Maintenance and repair order

- Goods movement document for material issue, receipt

- Man-hour confirmation

- Service entry sheet

- Billing document

6.2.2. Refurbishment Order Settlement

The settlement of a refurbishment order differs from that of a maintenance and repair order. A refurbishment order is settled against the material inventory (stock) of the subassembly material that has been refurbished.

During the refurbishment process, the order gets debited with the actual cost when the planned spare part and subassembly to be refurbished have been issued or the utilized man-hours have been confirmed. At the time of delivery (transferring the refurbished material to inventory), the order is credited with the equivalent value of the newly refurbished material (technical object). The credit value for the generated output is updated and displayed separately as a negative debit in the order.

At the time of settlement (see Figure 6-18), the balance credit transfer (difference between the order's debited cost and the generated output) from the order to the material inventory (stock) can be of two types.

- With the standard price control setting in the material master, the balance credit for the batch of the refurbished material is posted with the value of the standard price. The difference is posted as a credit memo to a general ledger account used for price differences in finance.

- With the moving average price control setting in the material master, the balance credit for the batch of the refurbished material is posted with the value of the actual increase in value, in other words, the value of the costs in man-hours and materials. This increases the moving average price at the material level.

The prerequisite customization for settling a refurbishment order to material inventory is that you have marked the material as the default account assignment in the settlement profile for the order type. This settlement profile is assigned to the refurbishment order type. At the time of refurbishment order processing, the settlement rule is automatically entered with the material as the settlement receiver (account assignment).

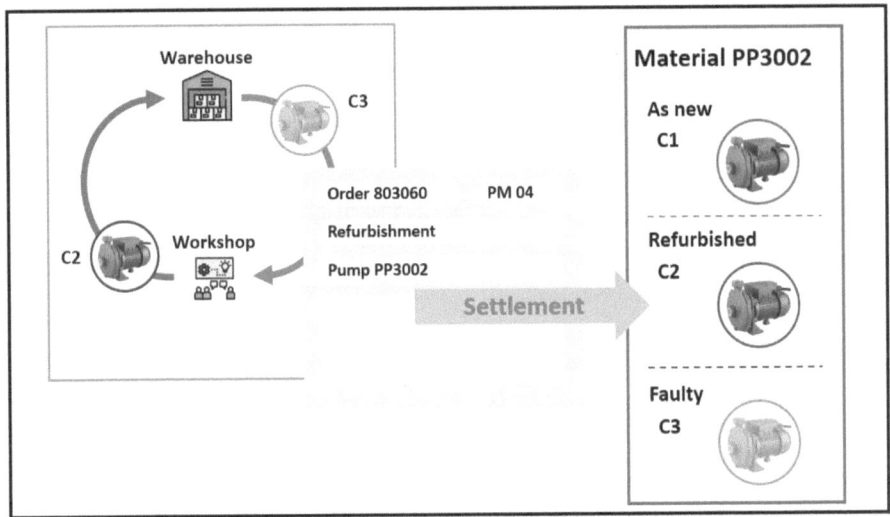

Figure 6-18. *Refurbishment order settlement*

6.2.3. Required Configuration Activities for Cost Settlement

Customization (configurations) related to an order's cost settlement involves the following.

- Settlement profile

- Allocation structure

- Time and creation of settlement/distribution rule

From the SAP Easy Access menu, navigate to Tools → Customizing. Double-click IMG → SPRO–Execute Project. Click the SAP Reference IMG button (see Figure 6-19).

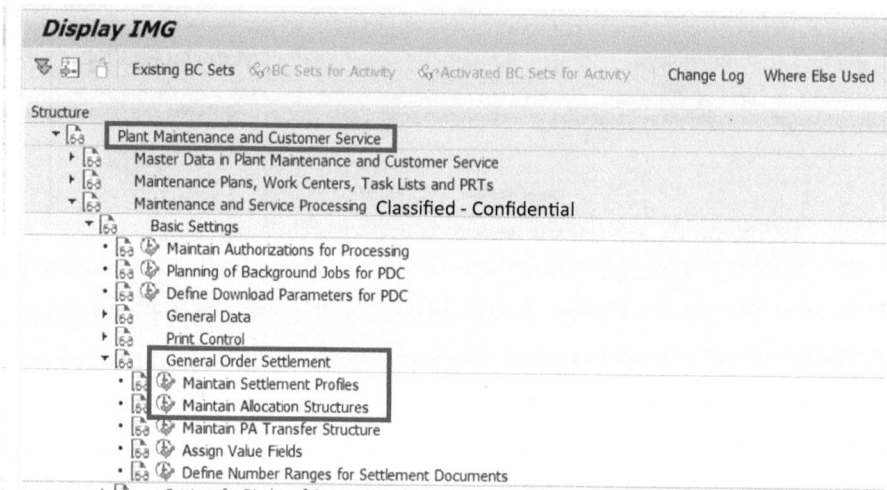

Figure 6-19. *Customization (configuration) nodes for maintenance order settlement*

Table 6-2 lists important configuration paths related to maintenance order settlement.

Table 6-2. *Configuration Path for Important Configuration Related to Maintenance Settlement*

Configuration Step	Configuration Path
Maintain settlement profile.	Plant Maintenance and Customer Service → Maintenance and Service Processing → Basic Settings → General Order Settlement → Maintain Settlement Profiles
Maintain allocation structure.	Plant Maintenance and Customer Service → Maintenance and Service Processing → Basic Settings → General Order Settlement → Maintain Allocation Structures
Configure order types (to assign settlement profile to order type).	Plant Maintenance and Customer Service → Maintenance and Service Processing → Maintenance and Service Orders → Functions and Settings for Order Types → Configure Order Types
Assign time and creation of settlement rule/distribution rule.	Plant Maintenance and Customer Service → Maintenance and Service Processing → Maintenance and Service Orders → Functions and Settings for Order Types → Settlement Rule: Define Time and Creation of Distribution Rule

6.3. Budgeting in Asset Maintenance and Repair

Budgeting the maintenance and repair of assets refers to setting, monitoring, and controlling the costs associated with performing large and high-cost maintenance activities on assets (equipment, machines, and assembly lines). Budgeting in maintenance orders helps organizations control maintenance expenses and make informed financial decisions (see Figure 6-20).

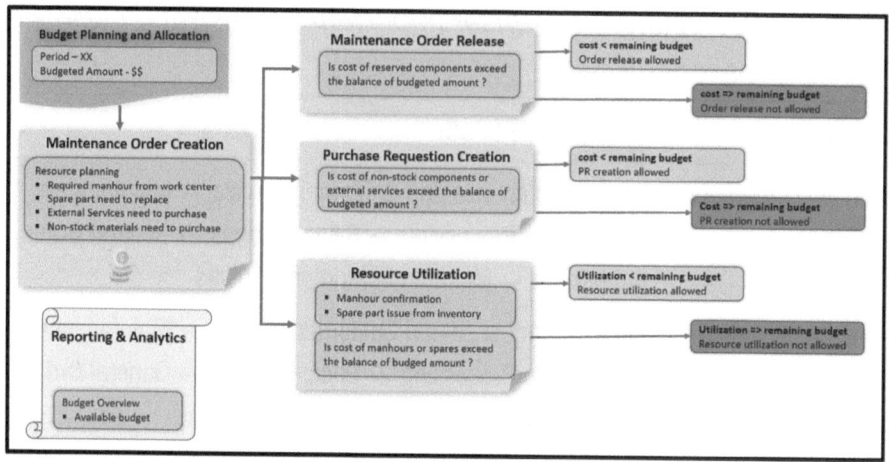

Figure 6-20. *Asset maintenance and repair budgeting*

Budgeting in asset maintenance and repair allows you to do the following.

- Plan and allocate budget as per the business requirement

- Monitor and control the consumption of spare parts, man-hours, and external services when the expense reaches the specified limit of the budget

- Analyze and make decisions based on available reports and analytics

Since asset management is seamlessly integrated with other S/4HANA business applications, budgeting in asset maintenance and repair processes can be set up in conjunction with the budgeting functionalities available in other business applications such as finance, project systems, and controlling. Budgeting in asset maintenance can be implemented in various ways, such as the following.

- Using funds management in the Finance module

- Using WBS in the Project System module

- Using the budget profile in the Controlling module

6.3.1. Budgeting Using Funds Management from the Finance Module

The Funds Management functionality serves to budget for all revenues and expenditures in individual functional areas of an enterprise, such as asset management. It controls future fund transactions according to the allocated budget and prevents exceeding it. The fund management functionality is integrated with the General Ledger Accounting component of the finance business application.

Integrating asset management and funds management applications allows you to oversee asset management processes that involve costs and require budgeting. As a prerequisite, integration between materials management and fund management is necessary for budget control during component issues for maintenance orders and the creation of external service and non-stock material purchase requisitions from maintenance orders. Additionally, integration with controlling and fund management (where FM-CO posting integration is active) is necessary for budget control during man-hour confirmation (confirmation of maintenance order operation with costing activity type).

The integration between asset management and funds management applications is established by maintaining an account assignment from funds management, such as commitment items, funds center, and funds, in the header section of a maintenance order.

During maintenance order processing and cost determination for planned resources, the system automatically selects the account assignment from the maintenance order.

The automatic determination of FM account assignment in a maintenance order occurs if you have entered the required assignments of FM account elements to CO objects in one of the steps of your derivation strategy. The system can extract the FM account elements from these assignments in this case. For instance, it can derive the FM account assignment from the requesting cost center entered in the technical asset (functional location, equipment master) of the maintenance order. The FM account assignment element determined through the account assignment derivation strategy appears as the default value and cannot be changed. Alternatively, the FM account assignment can be entered manually.

The following describes the process flow.

1. **Maintenance order creation**: When a maintenance order is created, the system determines and automatically populates an account assignment for funds management based on the defined derivation strategy. For manually managing the FM account assignment, navigate to Extras → Assignments → Funds Management in the maintenance order.

2. **Resource planning**: During the planning of resources for maintenance work, such as component reservation, purchase requisition for non-stock components, and external services, the system checks the available budget against the fund center linked to the cost center (such as the requesting cost center maintained in the technical asset) entered in the maintenance order's Account Assignment tab. If the cost of the required resources for the order reaches or exceeds the budget limit, the system prevents the release of the order and the generation of a purchase requisition.

3. **Operation confirmation**: During man-hour confirmation (confirmation of maintenance order operation with a costing activity type), the system checks the available budget. If the cost of the man-hours reaches or exceeds the budget limit, the system does not permit the saving of the completion confirmation.

Fund management is an application component within the finance and controlling module, integrating with various modules in S/4HANA, such as material management, project systems, asset management, and controlling. Users from the finance and controlling modules carry out all the necessary customizations. Therefore, customization activities are outside the scope of S/4HANA Asset Management and are not covered in this book.

6.3.2. Budgeting Using WBS in the Project System Module

Managing long-duration and high-cost maintenance work, such as shutdown maintenance, as a project enables you to allocate a specific budget to the project. This budget can be monitored and controlled, assisting you in keeping the costs in check and ensuring that the shutdown maintenance activities are completed within the allocated financial limits.

Integrating S/4HANA project system and asset management enables you to link maintenance orders with the WBS project. This integration facilitates budgeting in maintenance orders, as maintenance costs incurred during the order processing can be directly allocated to specific project WBS, aiding in better monitoring and controlling costs.

The user needs to calculate and allocate the budget to the respective WBS element to initiate budgeting in maintenance order processing. The WBS element with the allocated budget must then be entered in the maintenance order's WBS element field under the Additional Data tab.

During maintenance and repair planning for the maintenance order, the system compares the total planned cost for the required resources in the order with the remaining balance of the allocated budget for the WBS. The user can save the order if the budget limit is not reached. However, if the cost of the required resources for the order reaches or exceeds the budget limit (see Figure 6-21), the system prevents saving the maintenance order.

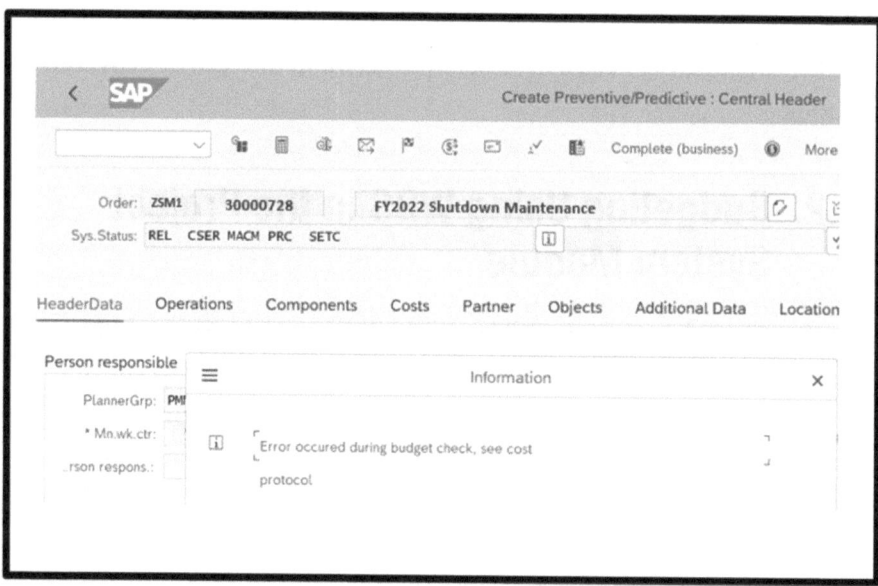

Figure 6-21. *Maintenance order not allowed to save when budget limit reached*

Budgeting using WBS does not require any specific customization, and typically, the existing setup of the project system module is equipped to start creating relevant master data, such as WBS elements, assigning budgets for the WBS elements, and defining tolerance limits for the budgeted value. After setting up the WBS element with a budget and tolerance limit, the WBS can be used in the maintenance order for budget control.

6.3.3. Budgeting Using Budget Profile from the Controlling Module

Another way of managing the budget for maintenance and repair orders is by using a budget profile from the controlling application. However, in this approach, the limitation is that you must allocate a budget to a specific maintenance order or a group of existing orders. This limitation restricts the usefulness of this approach.

Plan and Assign Budget

To initiate budgeting in maintenance order processing, the user must appropriately calculate and allocate the budget. In this approach to budget management, the budget can be allocated for a specific order or a group of existing orders (see Figure 6-22).

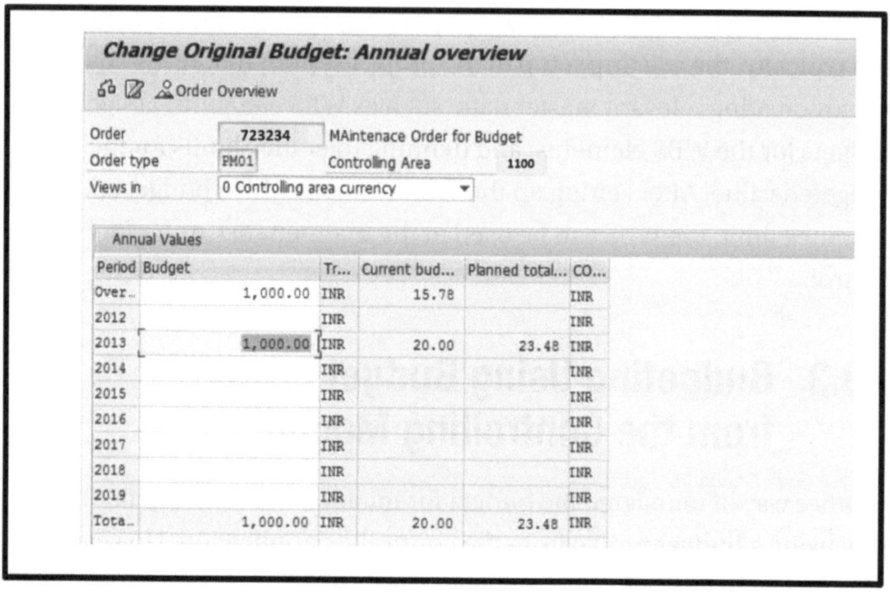

Change Original Budget: Annual overview

ᵔᵔ 🖉 👤 Order Overview

Order	723234	MAintenace Order for Budget	
Order type	PM01	Controlling Area	1100
Views in	0 Controlling area currency ▼		

Annual Values

Period	Budget	Tr...	Current bud...	Planned total...	CO...
Over..	1,000.00	INR	15.78		INR
2012		INR			INR
2013	1,000.00	INR	20.00	23.48	INR
2014		INR			INR
2015		INR			INR
2016		INR			INR
2017		INR			INR
2018		INR			INR
2019		INR			INR
Tota..	1,000.00	INR	20.00	23.48	INR

Figure 6-22. *Allocation of budget for maintenance orders or groups of orders for a year*

For example, a company wants to plan for a two-week-long shutdown maintenance work. The budget for annual shutdown maintenance has already been planned and released to the department.

For this work, the maintenance planner has created multiple orders. The planner has grouped all these orders and created an order group. The planned annual budget assigned for the order group. This budget applies to each order within the order group for the full year.

Maintenance and Repair Order Processing

During maintenance and repair planning for the orders from the group of orders, the system compares the total cost for the required resources with the remaining balance of the allocated budget for the group of orders. The user can process the order if the budget limit is not reached. However, if the cost of the required resources for the order reaches or exceeds the budget limit, the system does not allow the planned resources to be utilized (see Figure 6-23).

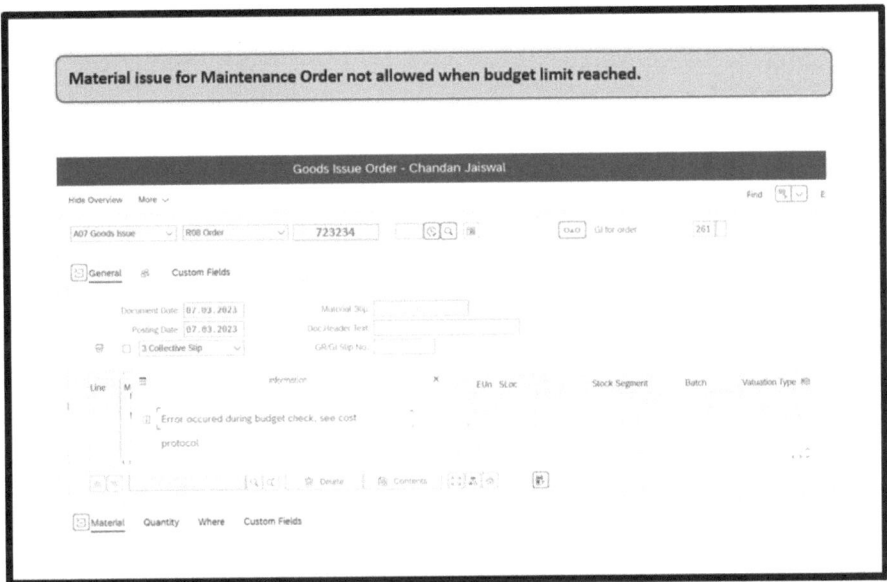

Figure 6-23. *Material issue for maintenance order not allowed when budget limit reached*

When the system does not allow the utilization of resources, the maintenance planner needs to review and replan the maintenance order (see Figure 6-24), aiming to reduce the required resources. This ensures that the maintenance cost does not exceed the allocated budget.

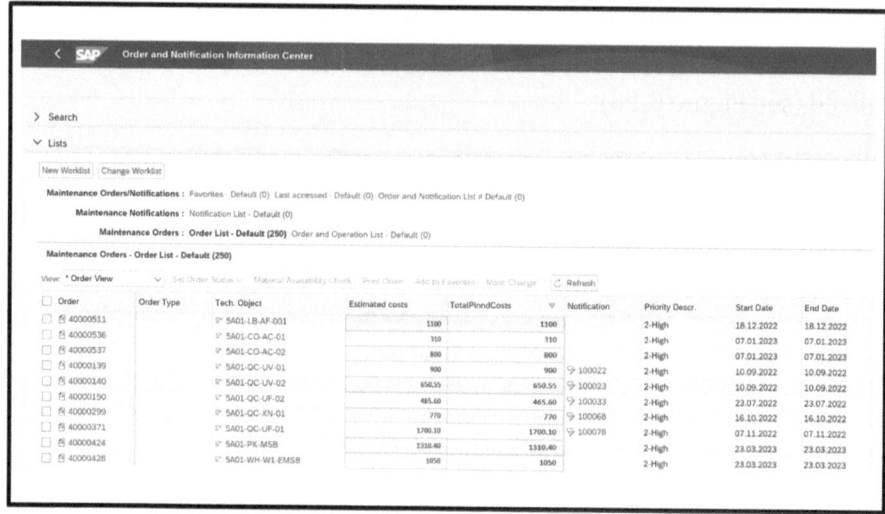

Figure 6-24. *Maintenance orders list with planned costs*

Setting up budgeting for maintenance orders involves the following customization steps.

1. **Create a budget profile.** To allocate a budget to maintenance order(s), you must define a budget profile. This profile includes information such as the time period for the validity of the budget and the activation settings for the budget profile, among other details.

2. **Define a tolerance limit.** In this customization step, you set up the tolerance limit for the budgeted value, such as 80%. This means the system triggers control measures when the order cost exceeds the allocated budget of 80% or more.

3. **Assign the budget profile to a maintenance order type.** In this customization step, you assign the budget profile to the maintenance order type for which budgeting is required.

Table 6-3 lists important configuration paths related to budgeting in maintenance order.

Table 6-3. *Configuration Path for Important Customization Related to Budgeting in Maintenance Order*

Configuration Step	Configuration Path
Maintain budget profile.	SAP Customizing Implementation Guide → Controlling → Internal Orders → Budgeting and Availability Control → Maintain Budget Profile
Define tolerance limits.	SAP Customizing Implementation Guide → Controlling → Internal Orders → Budgeting and Availability Control → Define Tolerance Limits for Availability Control
Assign a budget profile to the maintenance order type.	SAP Customizing Implementation Guide → Plant Maintenance and Customer Service → Maintenance and Service Processing → Maintenance and Service Orders → Functions and Settings for Order Types → Configure Order Types

6.4. Summary

The chapter delved into key aspects of managing costs and budgets in asset maintenance and repair processes. It covered costing in asset maintenance and repair, where integration with finance and controlling, account terms in maintenance orders, operation level costing, and costing value flow were explored. Additionally, topics such as overhead rates, commitments management, and costing in refurbishment of subassemblies were discussed with necessary configuration activities.

The chapter examined processes like maintenance order settlement, business completion, refurbishment order settlement, and the required configuration activities for cost settlement.

The chapter concluded with insights into budgeting in asset maintenance and repair, presenting various approaches to implementing budgeting in asset management.

This comprehensive chapter equips readers with valuable knowledge on effectively managing costs and budgets within asset maintenance and repair processes.

CHAPTER 7

Asset Management Integration with Other S/4HANA Business Applications

In the dynamic landscape of modern business applications, seamless integration between various applications is essential for optimizing processes and achieving operational excellence. This chapter delves into asset management integration within SAP S/4HANA and its interactions with other key business applications. By fostering a deep understanding of integration possibilities, benefits, and practical implementations, organizations can harness the full potential of their assets and drive efficiency across their operations. This chapter explores essential integration points and sheds light on their significance.

The following are some of the key topics covered in this chapter.

- Integration with S/4HANA supply chain applications

    - Materials Management

    - Production Planning

    - Quality Management

© Rajesh Ojha and Chandan Mohan Jaiswal 2023
R. Ojha and C. M. Jaiswal, *SAP S/4HANA Asset Management*,
https://doi.org/10.1007/978-1-4842-9870-1_7

- Integration with S/4HANA Financial Accounting and Controlling

- Integration with S/4HANA Project System

- Integration with S/4HANA Environment, Health, and Safety

Integrating modules in SAP S/4HANA means ensuring that different software parts work together smoothly. In S/4HANA, there are various modules (individual business applications), such as asset management, finance, inventory, sales, and human resources, and each module handles specific tasks. When you integrate these modules, it implies that they can share information and communicate with each other.

For example, there's a machine breakdown reported in the asset management module. When integrated with other modules (see Figure 7-1), like inventory management and purchasing, the system can automatically check if there are enough spare parts in stock for the maintenance job. If not, it can trigger a purchase requisition in the purchasing module to order the required parts.

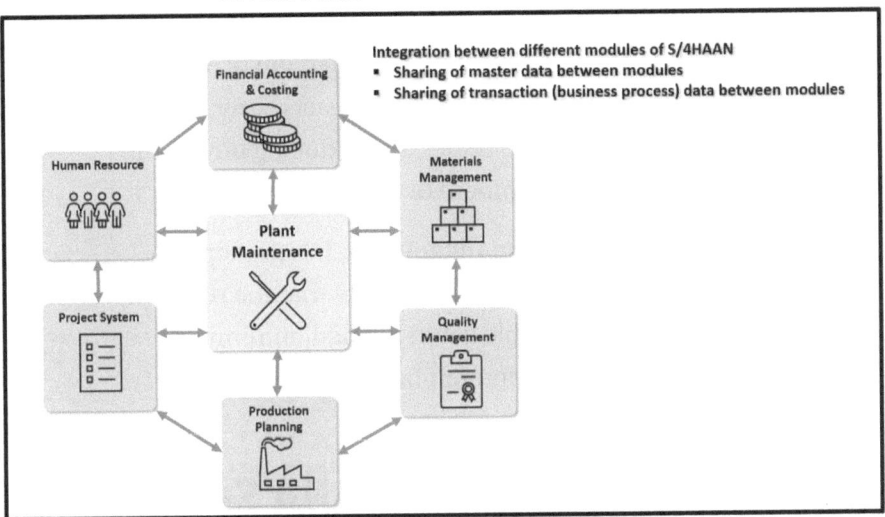

Figure 7-1. *Integration among different S/4HAAN business
applications*

The following explains some of the benefits of application integration.

- **Streamlined processes**: Integrating asset management
  with other modules in S/4HANA eliminates redundant
  data entry and manual handoffs between departments.
  This streamlines workflows, reducing errors and
  improving overall efficiency.

- **Real-time visibility**: Integration provides real-time
  data exchange between modules, allowing better
  visibility into material availability, quality inspection
  results, production schedules, and financial and
  costing reports. This enables quicker decision-making
  and better resource planning.

315

- **Reduced downtime**: Integration with quality management allows immediate inspection of materials and products. This helps identify defects early, preventing production issues and reducing downtime caused by faulty equipment or materials.

- **Optimal spare parts management**: Integration with Materials Management ensures that the right spare parts are available when needed, minimizing downtime due to equipment breakdowns.

7.1. Integration with S/4HANA Supply Chain Applications

S/4HANA Asset Management helps businesses manage their asset maintenance process efficiently. When integrated with other S/4HANA Supply Chain modules, like Materials Management (MM), Quality Management (QM), and Production Planning (PP), it forms a comprehensive solution that optimizes the entire maintenance process by sharing data among these modules.

7.1.1. Materials Management

S/4 HANA Asset Management and Materials Management (procurement and inventory) are two essential modules in the S/4HANA Business Suite that help businesses maintain their assets and manage their material effectively. Next, let's discuss a few important integrations between these two modules.

Initiating a Materials Purchase Requisition from a Maintenance Order

For the maintenance and repair of an asset, sometimes you need materials
that are not part of the regular inventory (non-stock materials). For example,
it could be a specialized high-cost spare part or a one-time purchase item.
If you realize that these non-stock materials are required when planning a
maintenance order, you can automatically create a purchase requisition by
entering the non-stock material details, such as the material master number,
required quantity, and issuing plant and releasing the maintenance order.
Releasing the order automatically creates a purchase requisition for the
required material and quantity. The purchasing department then takes your
purchase requisition and proceeds with the purchase of the material by
converting the purchase requisition to a purchase order. When the vendor
supplies the material, the maintenance order is debited with the cost of the
material upon goods receipt in the Materials Management module.

Initiating an External Service Purchase Requisition from a Maintenance Order

Similar to the purchase requisition for non-stock materials, if you need any
specialized workmanship (such as the calibration of an electronic system
or specialized welding) from an external supplier, you can automatically
create a service purchase requisition.

The maintenance order provides two types of procurement processes
for the processing of an operation externally by an external vendor.

Procurement of Service

For a maintenance order operation that needs to be processed by an
external vendor, and where the vendor delivers the required service at the
customer's location where the technical asset is maintained (meaning
the technical asset is not required to be sent to the vendor's location,

and therefore, goods movement of the technical asset is not required),
you need to select control key PM03 for the operation. For example, the
calibration of a large weighing scale needs to be done by an external
vendor, and the vendor performs the calibration at the location where the
weighing scale has been commissioned.

An operation with a PM03 control key is not required to be scheduled,
and capacity requirements are not generated for the maintenance work
center. After entering the control key PM03 and other purchasing-related
details (e.g., the service master number, required quantity, and releasing
the maintenance order), a purchase requisition is automatically created.
The purchasing department then takes your purchase requisition and
proceeds with the procurement of external services. A service entry must
be performed in the Materials Management application to confirm the
operation's completion.

Subcontracting for an Operation

If a maintenance order operation needs to be processed by an external
vendor through subcontracting, wherein the technical object may need to
be sent to the external vendor's location, then you must select control key
PM02. For example, a large electrical motor requires rewinding, and for
this, the technical asset housing the motor needs to be sent to the vendor's
location since the machines and tools used for rewinding are installed at
the vendor's location. Goods movement of the technical asset is required
so that the technical asset can be issued to the vendor.

An operation with a PM02 control key can be scheduled; capacity
requirements are generated. A purchase requisition is automatically
created after entering the PM02 control key and releasing the maintenance
order. The purchasing department then takes your purchase requisition
and proceeds with the procurement process. After the completion of the
required activity by the external vendor, a goods receipt needs to be posted
in Materials Management to finalize the delivery of the purchase order.

Reserving, Issuing, and Costing of Inventory Material from a Maintenance Order

For the maintenance of an asset, while planning a maintenance order, you can reserve all the required materials (spare parts) to carry out maintenance operations (tasks). Upon releasing the order after planning, the system automatically creates reservations for the planned materials in the Materials Management module. On the Materials Management module's inventory side, the planned materials' required quantity is reserved for the maintenance order (see Figure 7-2). This implies that a requirement for the necessary material and quantity is generated, and both the warehouse clerk and any other requester can view the total available quantity and total reserved quantity for the material. Upon issuing the planned materials to the maintenance order, the order is debited with the cost of the material. Additionally, as part of the financial accounting, the corresponding G/L account entries (debit and credit) happen automatically.

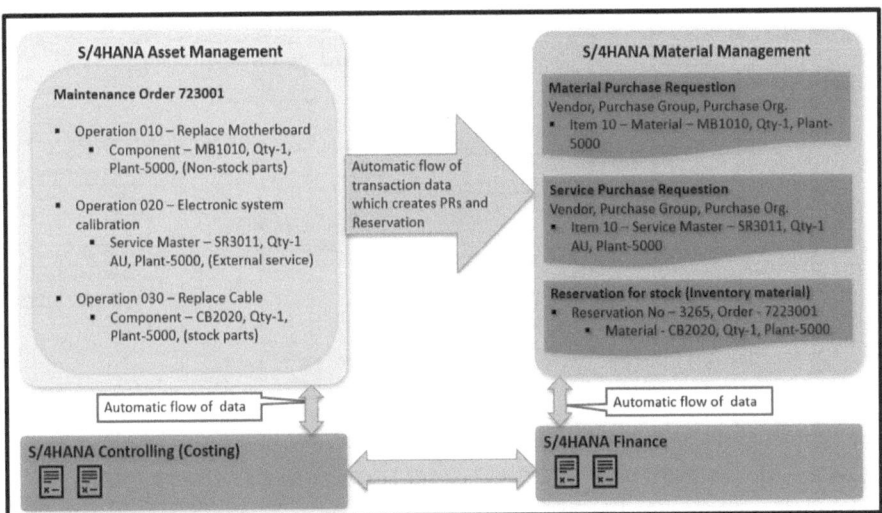

Figure 7-2. *Asset Management and Materials Management integration*

7.1.2. Production Planning

Generally, assets (such as machines, equipment, and assembly lines) are
employed in the manufacturing process by using them as production work
centers. Consequently, if a maintenance order is created and released
for carrying out maintenance work for an asset, the production planner
should have the ability to view all released maintenance orders in their
planning report (e.g., planning board) for those production orders where
the production work center consists of the under-maintenance asset.

To accomplish this, the production work center code must be entered
in the master data of the technical asset (equipment master's location data
section).

In the header section of the open maintenance and repair order, the
"0 – Not in operation" value in the System Condition field must be selected
(see Figure 7-3). The maintenance order must be released for execution.

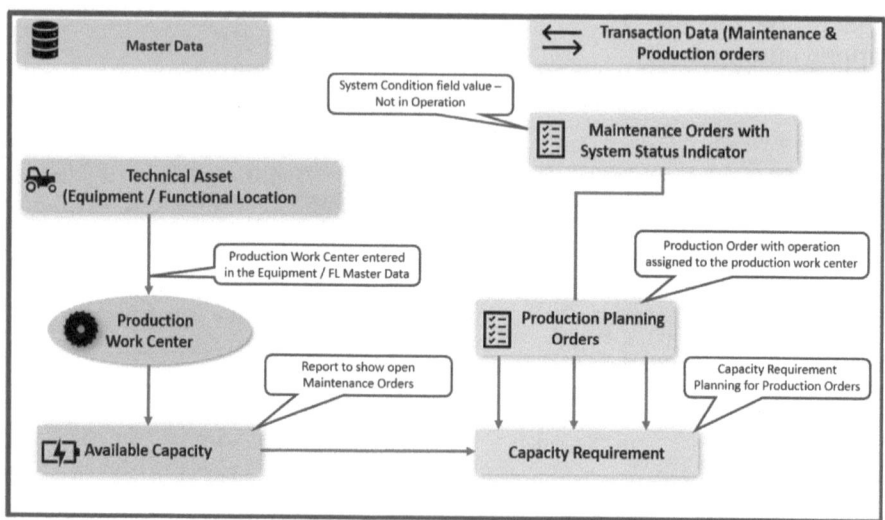

Figure 7-3. *Integration between asset (equipment master) and
production capacity planning*

The capacity planning report shows the maintenance order (see Figure 7-4). When the production planner tries to dispatch the production order per the planned schedule, it is moved to the next/previously available capacity based on the scheduling strategy. If the production department needs to use this planned schedule for production, the maintenance order must be moved out of this planned timeline. This can be done by rescheduling the maintenance order.

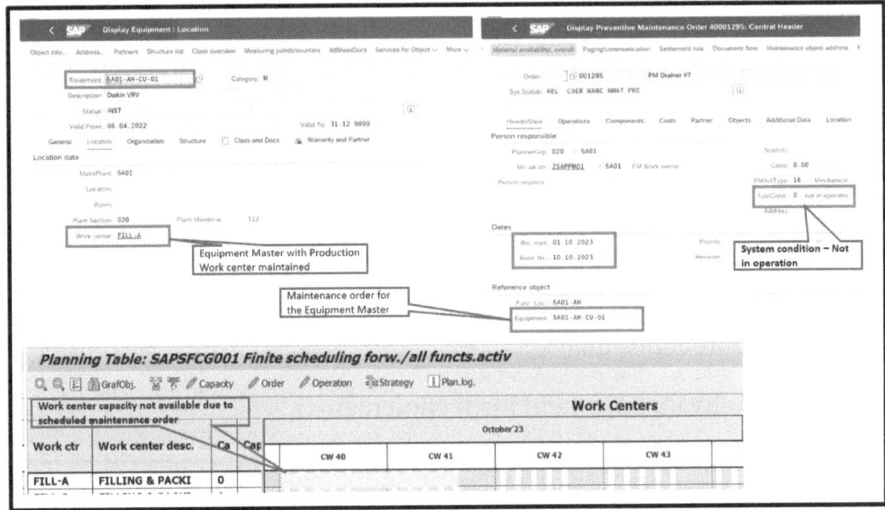

Figure 7-4. *Production work center capacity is not available due to scheduled maintenance order*

7.1.3. Quality Management

Inspecting an asset based on the recommended schedule is required to ensure that the asset meets the quality standard mandated by quality control, which is legally binding. Asset management is integrated with quality management and supports the inspection checklist process, from creating the inspection plan to the resulting recording and follow-up actions in asset management.

To plan the inspection of assets, maintenance planning, and
scheduling functionalities are being used to generate maintenance
orders regularly as per the required schedule. The inspection checklist
(inspection lot) in quality management is generated and assigned to the
maintenance order based on

1. The classification data is maintained for the
    technical asset and inspection plan.

2. The checklist type is assigned to the operation of a
    maintenance order

The following lists important master data (see Figure 7-5) for the
inspection checklist process.

-   Equipment master, functional location (technical asset)

-   Task list

-   Maintenance plan

-   Quality management inspection plan

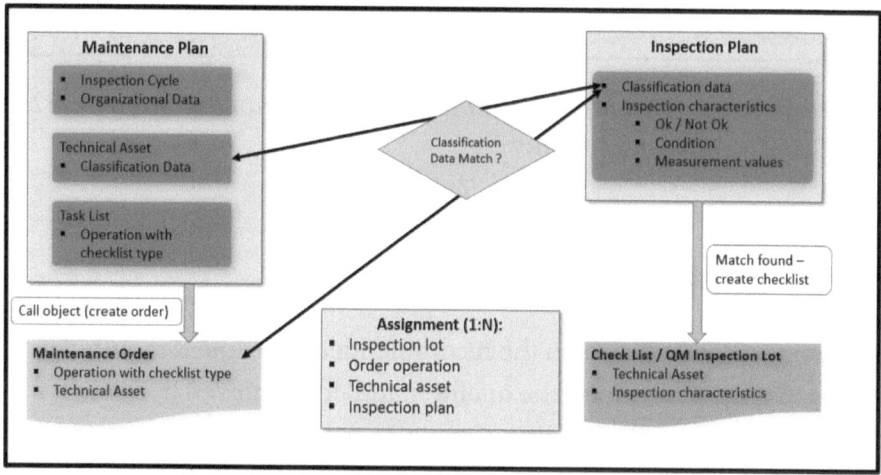

Figure 7-5. *Inspection checklist master data*

Process Flow

Each inspection job is initiated with a maintenance order created using a
maintenance plan. The characteristics data of classification maintained in
each inspection plan is compared with the classification of the technical
asset and the checklist type of an operation in the maintenance order. The
complete inspection checklist process comprises the following.

1. Create a maintenance order from the maintenance
   plan (or manually for ad hoc inspection).

2. Generate object list (automatically or manually).

3. Create an inspection checklist based on the
   object list.

4. Record results for inspection lot (checklist items)
   based on inspection lot characteristics.

5. Select a usage decision for the inspection list.

6. Confirm resources and complete
   maintenance order.

7. Continue with follow-up actions after the usage
   decision for the inspection lot.

Object Lists

At the time of maintenance order generation or after the order is created,
an object list is generated from the header section of the maintenance
order or objects from the maintenance plan item. Subsequently, during
inspection checklist generation, each technical object (equipment,
functional location) in the object list is checked to find a matching
inspection plan.

Inspection Checklist and Inspection Lot

After the object list is generated and inspection plans for each technical object have been found, a new inspection lot is created, which can be viewed in the Checklists tab.

Result Recording and Usage Decision for Inspection Checklist

The user needs to record the inspection result based on the inspection lot's characteristics. You can set a usage decision based on the inspected results. When you set a usage decision, all follow-up actions defined are executed for the selected inspection lot and maintenance order.

7.2. Integration with S/4HANA Financial Accounting and Controlling

Integration between S/4HANA Finance, Controlling, and Asset Management ensures a smooth data flow and coordination between financial accounting and maintenance management. It helps maintain accurate asset and equipment information, enables activity-based costing for maintenance operations, and allows proper settlement of incurred costs to relevant financial destinations. This integration ultimately enhances an organization's overall efficiency of financial and maintenance processes. The following are important integrations among these modules.

- Asset master and equipment master synchronization

- Costing value flow in maintenance order

- Maintenance order cost settlement to cost object

7.2.1. Asset Master and Equipment Master Synchronization

In the S/4HANA Finance application, the asset master is created in relation to specific equipment or functional locations in the asset management module. Depending on the organization's process regarding the price value of a technical object (equipment/machines/assembly lines), the object is created as an asset master in finance alongside the creation of the equipment master or functional location in the plant maintenance (asset management) module. Since both the equipment master and asset master represent a single technical object, it is crucial to maintain accurate and consistent master data for reliable business transactions and reports on these data.

The main difference between asset master and equipment master is that the asset master is a central record that contains detailed information about important assets owned by a company, such as machinery, vehicles, or buildings. It serves as a single source of truth for all asset-related data, including purchase information, depreciation values, and useful life. On the other hand, the equipment master in plant maintenance (asset management) stores essential information about specific equipment or machinery that needs maintenance and monitoring. It includes details like equipment type, technical specifications, and maintenance history.

Synchronization between the asset master and equipment master ensures that when you create or update information about an asset in the asset master, it automatically synchronizes with the corresponding equipment master in asset management. The synchronization can also be set up in reverse order so that whenever an equipment master in asset management is created or changed, the corresponding asset master is created or changed respectively in finance's asset accounting. This synchronization helps maintain consistency and accuracy between financial data (asset details) and maintenance data (equipment details) throughout the system.

Information synchronized in the master data includes company code, plant, manufacturer, and cost center. To link the asset master and equipment master for synchronization (see Figure 7-6), the asset master number is maintained in the Asset field in the Organization tab of the equipment master.

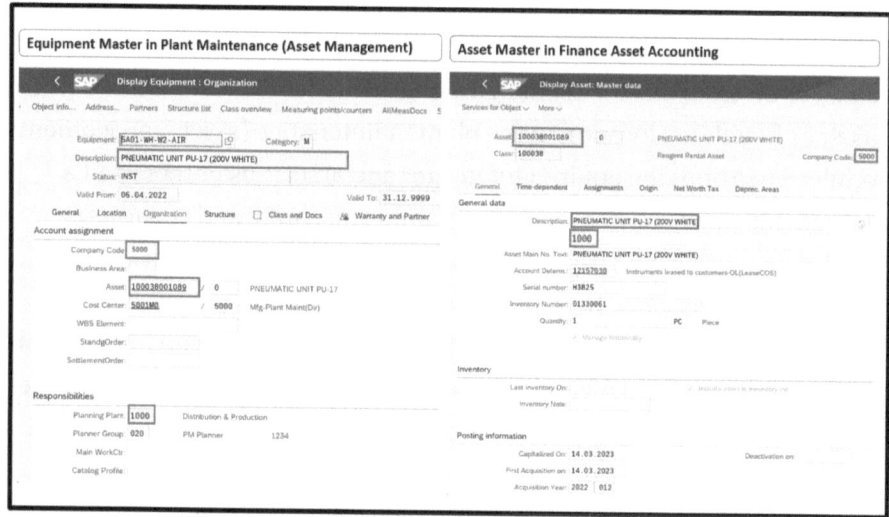

Figure 7-6. *Asset master and equipment master synchronization*

7.2.2. Costing Value Flow in Maintenance Order

Costing in maintenance orders is a way to track and manage the expenses related to maintenance activities in a business. It helps organizations keep a close eye on the costs incurred during the maintenance process (see Figure 7-7), such as servicing equipment, repairing machinery, or any other maintenance-related activities.

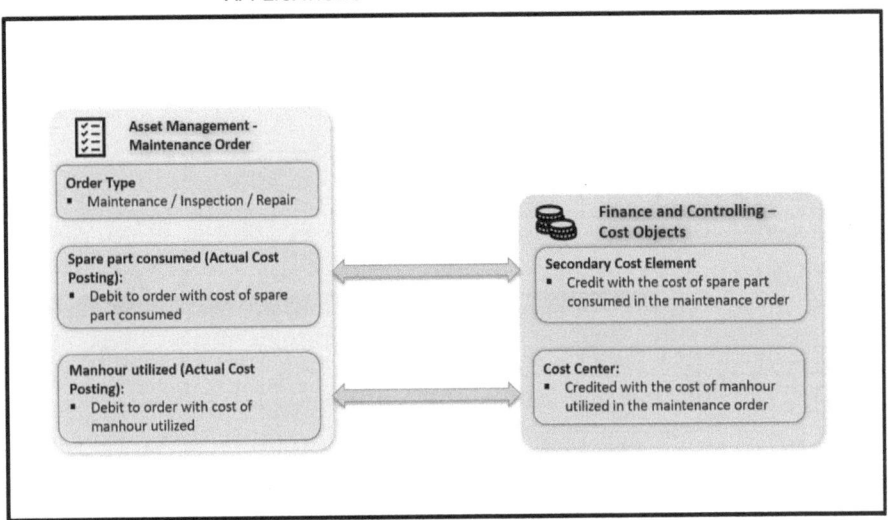

Figure 7-7. *Costing value flow between maintenance order and
cost objects*

During maintenance, inspection, and repair of a technical object using
a maintenance order in asset management, the order gets debited with
planned costs when the required resources, such as spare parts and man-
hours, have been planned, and the order has been released (dispatched).
The order is debited again with the actual cost when the planned spare
part has been consumed, or the utilized man-hours have been confirmed
(see Figure 7-8).

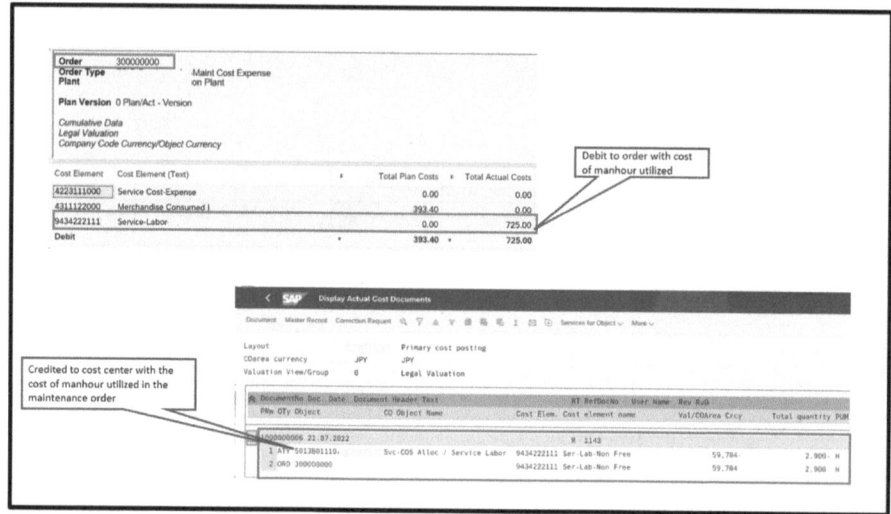

Figure 7-8. *Costing value flow between maintenance order and cost objects*

7.2.3. Maintenance Order Cost Settlement to Cost Object

Settlement is a cost allocation process, fully or partially, from one cost object to another. Settling the maintenance and repair order transfers the costs from the order to the relevant cost centers or other cost objects (see Figure 7-9) where these expenses belong. This way, the financial reports become accurate, and you can better understand the true financial impact of the maintenance activities related to asset management.

During the completion phase of the maintenance order, the order is settled to an actual cost receiver, which means the order is credited by allocating its debit to the receiving cost object.

For example, in a manufacturing industry, the settlement receiver can be a cost center of the repaired technical object. This means the maintenance order collects all the resource costs for reporting and analysis at the maintenance order level. Subsequently, the cost is transferred to receiving cost objects where these expenses belong.

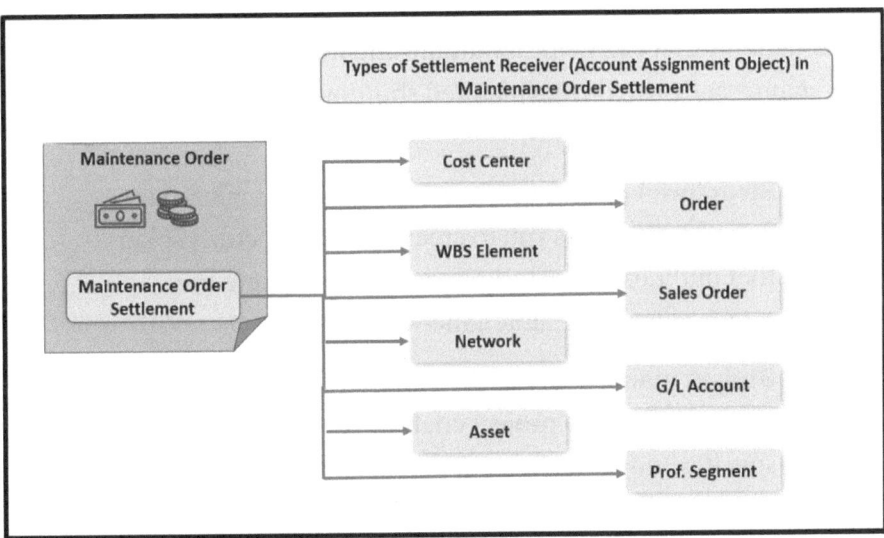

Figure 7-9. *Various types of cost settlement receivers in maintenance order settlement*

7.3. Integration with S/4HANA Project System

Project System (PS) is like a smart organizer for big tasks. It helps companies plan, manage, and track projects, like building a new factory or launching a product. It keeps all the details in one place, so everyone knows what's going on, what needs to be done, and how much it's costing.

Integrating the Project System and Asset Management applications
allow you to link maintenance orders with the work breakdown structure
(WBS) project. This integration enables accurate cost tracking, as
maintenance costs incurred during the shutdown can be directly allocated
to specific WBS projects, helping in better cost analysis and reporting.

Organizations treat large and long-duration maintenance jobs, such
as shutdown maintenance, as a project in Project System because it offers
several benefits that can help streamline and improve the management of
maintenance activities during planned shutdowns. The following are some
of the key advantages.

- **Structured planning**: Treating shutdown maintenance
  as a project allows you to create a well-structured plan
  that outlines all the tasks, resources, and timelines
  required for the maintenance activities.

- **Budget management**: Managing shutdown
  maintenance as a project enables you to allocate a
  specific budget to the project. This budget can be
  monitored and controlled, helping you keep the costs
  in check and ensuring that the maintenance activities
  are completed within the allocated financial limits.

7.3.1. Settlement of Maintenance Order to WBS in Project System

You need to maintain the WBS as an account assignment in the Location
tab of the maintenance order and as a settlement receiver in the settlement
rule of the order. Once the order execution is complete, the order is settled
to the WBS. All the costs in the maintenance order are transferred to the
WBS (see Figure 7-10). This enables tracking and managing maintenance
costs within your larger project, using the WBS as a connecting point.

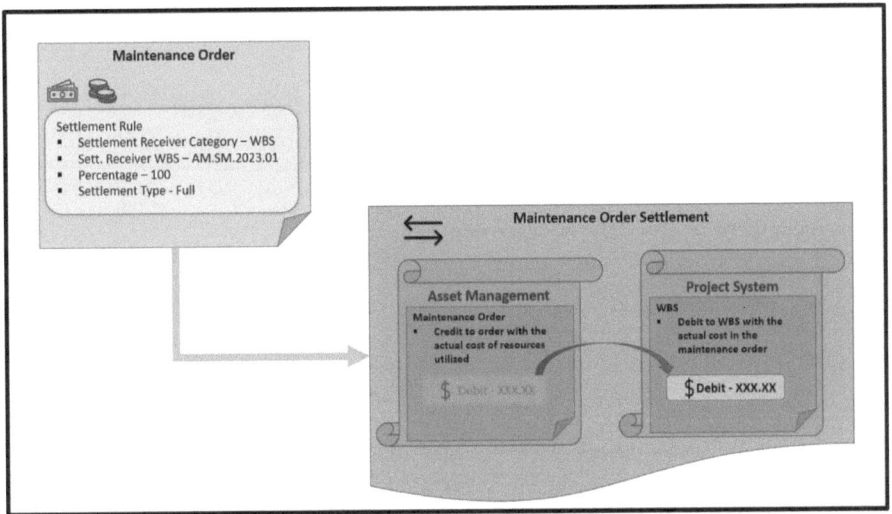

Figure 7-10. *Settlement of maintenance order to WBS in
Project System*

The prerequisite customization to allow WBS as a settlement receiver
in a maintenance order is to set the settlement receiver category as WBS
optional (see Figure 7-11) in the settlement profile used for the order type.

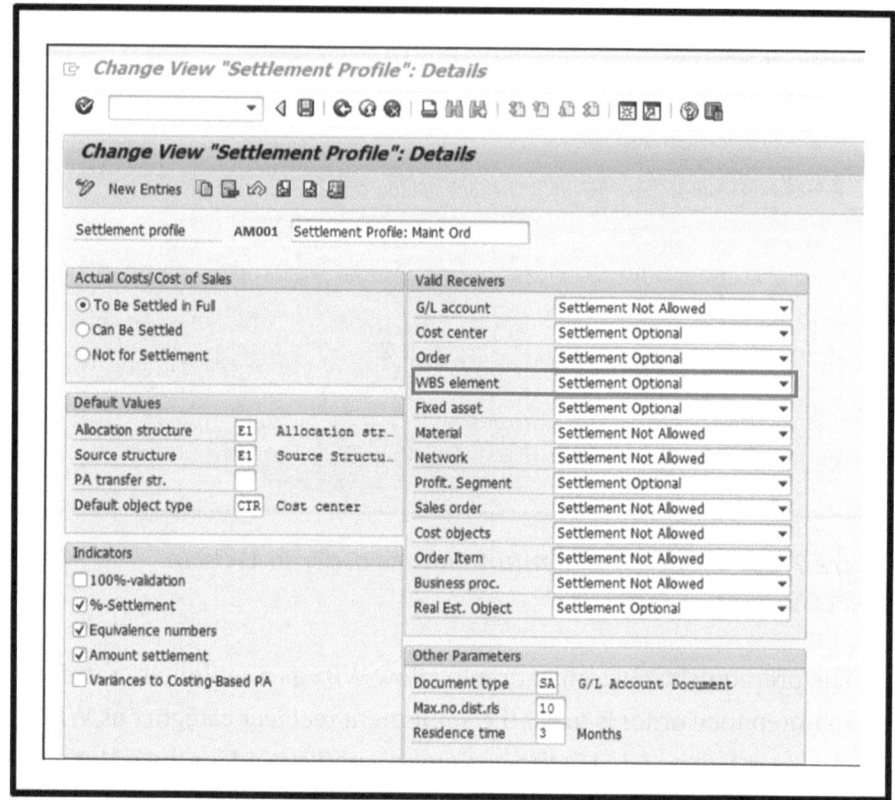

Figure 7-11. Settlement receiver is WBS optional in settlement profile

7.3.2. Budgeting for Maintenance Order Using the WBS in Project System

Another integration possibility between the Asset Management and
Project System applications is budgeting for maintenance orders using
the WBS.

Budgeting in the maintenance and repair of assets refers to the
process of setting, monitoring, and controlling the costs associated
with performing large and high-cost maintenance activities on assets
(equipment/machines/assembly lines). It helps plan and control

expenses related to maintenance activities. At the start of maintenance order processing, the user must calculate and allocate the budget to the respective WBS element. The WBS with the allocated budget must be maintained in the "WBS element" field under the Additional Data tab of the maintenance order.

During maintenance and repair planning for the maintenance order, the system compares the total planned cost for the required resources in the order with the remaining balance of the allocated budget for the WBS. The user can save the order if the budget limit is not reached. However, if the cost of the required resources for the order reaches or exceeds the budget limit (see Figure 7-12), the system does not allow you to save the maintenance order.

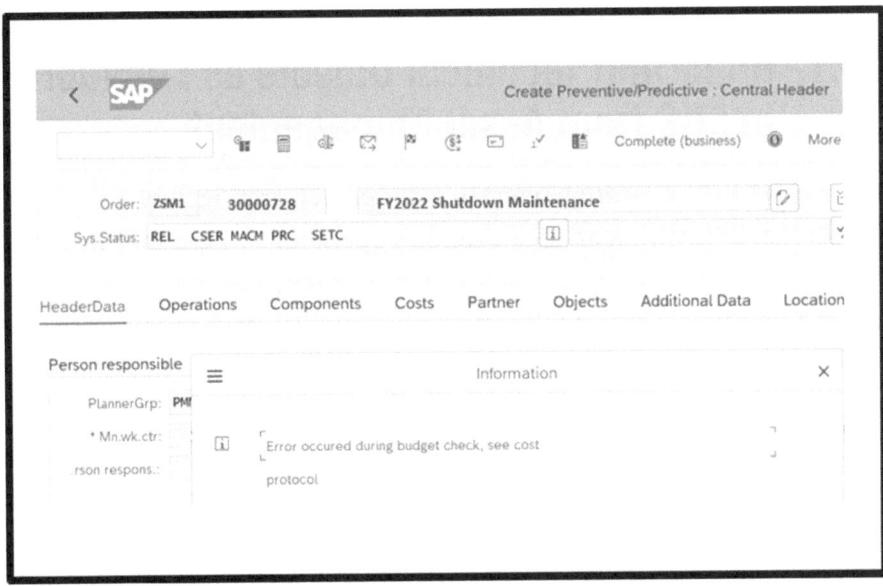

Figure 7-12. *Maintenance order not allowed to save when budget limit reached*

7.4. Integration with S/4HANA Environment, Health, and Safety (EHS)

S/4HANA EHS deals with environment, health, and safety. It's an
application module that helps companies take care of essential business
processes, such as ensuring a safe environment, maintaining employee
health, and adhering to regulations to protect everyone. This module aids
businesses in managing aspects like safety procedures, health regulations,
and environmental responsibilities, ensuring they take appropriate
measures to keep everyone safe and well.

EHS and Asset Management can be integrated if required. Next, let's
discuss the two key integrations between these two modules.

7.4.1. Importing Technical Objects as Locations in EHS from Asset Management

In S/4HANA EHS, a *location* refers to a designated area, facility, or
geographical site within an organization's operations where specific EHS-
related activities, processes, or incidents occur. It serves as a categorization
and organizational tool, enabling the linkage of EHS data and processes
to particular physical or virtual spaces. This aids in effective tracking,
reporting, and compliance management.

Technical objects such as equipment masters and functional locations
from asset management (plant maintenance) can be imported and used
as locations in EHS. The advantage of this is that you do not have to
enter duplicate data; the foundational data for these locations originates
from asset management. Alternatively, you can link existing locations
in EHS Management with technical objects from asset management.
Consequently, the system inserts the location into the location structure,
making it available for various business processes within the application,
such as reporting incidents (see Figure 7-13).

The following are prerequisites.

- Enable Asset Management integration in Customizing
  for Incident Management or Risk Assessment in
  S/4HANA EHS Management.

- Specify location types in Customizing for Incident
  Management or Risk Assessment in S/4HANA EHS
  Management.

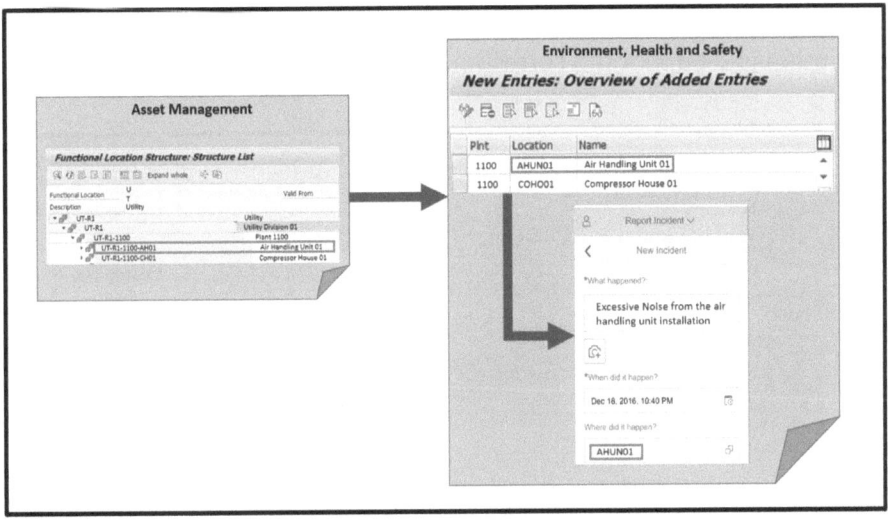

Figure 7-13. *Importing EHS locations from Asset Management*

7.4.2. Triggering Maintenance Notification from S/4HANA EHS Incident Management

The incident functionality in EHS supports you in capturing, processing,
and completing events related to work. Here, an event represents
everything you must capture or wish to record, such as personal injuries,
accidents, damage to facilities, and any abnormalities-related findings.

During incident creation, you can initiate maintenance notifications
within the incident management to request the maintenance and repair
work required to prevent and rectify an unsafe condition. The user needs
to select a notification type. The system assigns a notification ID and adds
the maintenance notification as a new process in the Tasks table, and the
process appears in the Task Processes table.

For example, a maintenance notification (see Figure 7-14) is created to
investigate the root cause of excessive noise from an air handling unit and
resolve the problem to stop noise pollution and other risks.

As a prerequisite, if you wish to add notifications, ensure that
integration is set up for the corresponding systems.

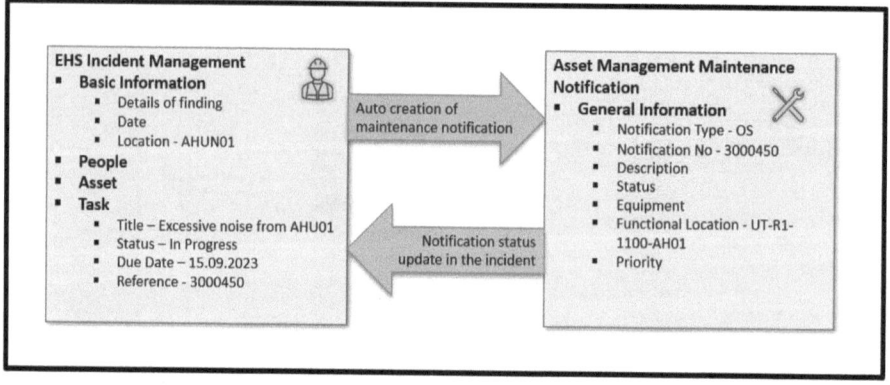

Figure 7-14. *Maintenance notification generation from EHS incident
management*

7.5. Summary

This chapter delved into the crucial integration of Asset Management
with various other S/4HANA business applications. It underscored the
paramount importance of seamless integration, showcasing how these
connections enhance operational efficiency, cost management, and
overall business performance.

CHAPTER 8

Innovation with Asset Management

This chapter discusses master data—the foundational elements crucial for managing assets effectively—in SAP Enterprise Asset Management.

8.1. SAP Intelligent Asset Management Suite

SAP Intelligent Asset Management (IAM) is a comprehensive solution that leverages digital technologies to optimize the management of physical assets within an organization. It encompasses a set of tools, applications, and capabilities designed to enhance asset performance, reduce downtime, improve maintenance practices, and drive operational efficiency.

SAP IAM constitutes an essential element of the SAP Intelligent Enterprise framework. Within the realm of SAP IAM, SAP's interconnected digital twin ecosystem synchronizes the virtual, physical, condition-related, and commercial definitions of assets and products in real time. This synchronization expedites innovation, optimizes operational performance, anticipates service needs, enhances diagnostic capabilities, and elevates decision-making quality.

© Rajesh Ojha and Chandan Mohan Jaiswal 2023
R. Ojha and C. M. Jaiswal, *SAP S/4HANA Asset Management*,
https://doi.org/10.1007/978-1-4842-9870-1_8

This network of digital twins facilitates a cooperative end-to-end digital transformation spanning the stages: design and manufacturing, operations and maintenance, and service and end-of-life management. This, in turn, supports the following.

- The seamless exchange of digital twin data across all organizational departments

- Collaborative interaction with suppliers, customers, and service providers throughout the entire lifecycle

- Pioneering innovative business models for products and services

You see the following advantages on the enterprise level.

- Leverages data assets for expedited achievement of desired outcomes with reduced risk

- Enables employees through the automation of processes

- Foresees and takes initiative-taking actions to address customer requirements

- Innovates novel business models and sources of revenue

The SAP IAM portfolio (see Figure 8-1) encompasses a range of deployment options, including cloud-based, on-premise, and mobile solutions. This variety of deployment models caters to diverse customer preferences.

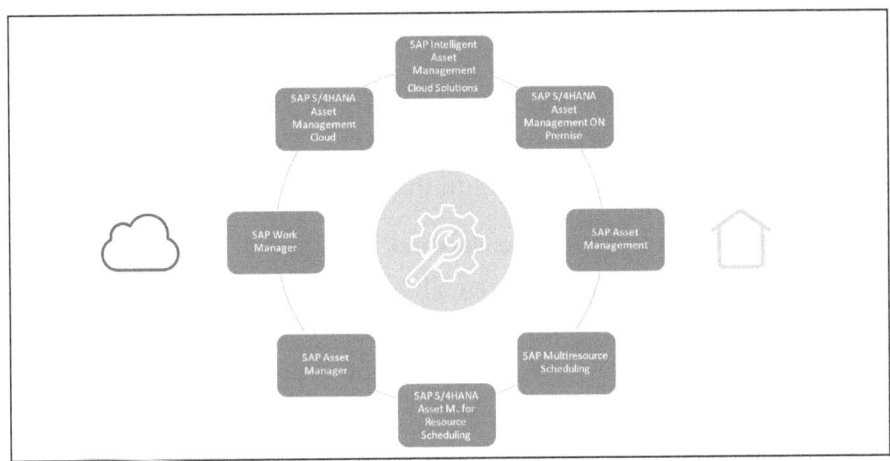

Figure 8-1. *SAP IAM portfolio*

The following is included in the portfolio.

- SAP Cloud Solutions for Intelligent Asset Management

- SAP S/4HANA Cloud Asset Management

- SAP S/4HANA On-Premise Asset Management

- SAP ERP Asset Management

- SAP Asset Manager

- SAP Work Manager

- SAP Multiresource Scheduling

- SAP S/4HANA Asset Management for Resource Scheduling

8.2. SAP IAM Cloud Solutions

SAP IAM portfolio includes cloud-based solutions that aim to enhance asset management practices using intelligent technologies. Figure 8-2 illustrates some SAP IAM Cloud Solutions.

- **SAP Asset Intelligence Network**: This cloud-based solution provides a collaborative platform for connecting manufacturers, operators, and maintenance personnel. It offers a digital representation of assets, enabling stakeholders to share and access information about asset specifications, documentation, and maintenance requirements.

- **SAP Predictive Asset Insights**: This solution utilizes predictive analytics and machine learning to forecast equipment failures and suggest maintenance actions before issues arise. It helps organizations optimize maintenance schedules and reduce unplanned downtime.

- **SAP Asset Performance Management**: This solution assists in developing asset management strategies based on historical data, performance goals, and industry best practices. It helps align maintenance strategies with business objectives and optimize asset performance.

- **SAP Asset Central**: This solution provides a centralized platform for managing asset information, maintenance records, and documentation. It aims to improve collaboration among maintenance teams and streamline maintenance processes.

- **SAP Mobile Asset Management**: While not exclusively part of the IAM portfolio, this cloud-based solution integrates with SAP Intelligent Asset Management to manage field service operations efficiently. Features include scheduling, resource allocation, and mobile access for field technicians.

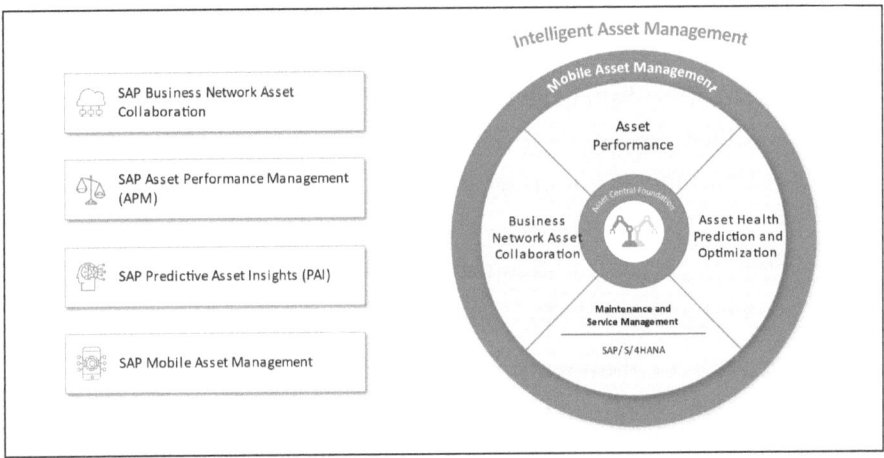

Figure 8-2. *SAP IAM overview*

8.3. SAP Business Network Asset Collaboration

SAP Business Network Asset Collaboration focuses on leveraging digital platforms and networks to facilitate collaboration and information exchange among various stakeholders involved in the lifecycle of assets. This collaborative approach connects manufacturers, suppliers, operators, maintenance teams, service providers, and other relevant parties through a shared digital ecosystem. The goal is to enhance visibility, communication, and coordination throughout the asset's lifecycle, from design and manufacturing to operation, maintenance, and even end-of-life stages.

The following are key features of SAP Business Network Asset Collaboration.

- **Digital twin sharing**: A digital twin is a virtual representation of a physical asset or product. In asset collaboration, stakeholders can share and access digital twin data, specifications, documentation, and performance information. This fosters a collective understanding of the asset's characteristics and requirements.

- **Real-time data exchange**: Through connected sensors and IoT devices, real-time data about asset conditions, performance, and usage can be collected and shared among relevant parties. This enables initiative-taking decision-making and timely responses to maintenance needs.

- **Supplier collaboration**: Manufacturers and suppliers can collaborate to ensure that asset components and parts are high quality and meet specifications. Supplier collaboration can also involve sharing maintenance and service information for the assets they provide.

- **Maintenance and service coordination**: Maintenance teams, both in-house and external service providers, can collaborate to optimize maintenance schedules, share diagnostic data, and execute maintenance tasks more efficiently.

- **Lifecycle transparency**: All stakeholders gain transparency into the asset's lifecycle stages, including design, production, installation, operation, maintenance, and eventual retirement or replacement. This transparency aids in making informed decisions at every stage.

- **Risk mitigation and compliance**: Collaborative asset management can help organizations ensure compliance with regulatory requirements and safety standards. It also enables identifying and mitigating potential risks associated with asset operation.

- **Innovation and continuous improvement**: Collaboration can lead to innovative solutions and continuous improvement efforts as different stakeholders bring their expertise to the table. This can result in enhanced asset performance and extended asset lifecycles.

Business Network Asset Collaboration often involves using digital platforms, cloud-based solutions, and emerging technologies such as the IoT, artificial intelligence (AI), and data analytics. By bringing together various stakeholders and data sources, organizations can optimize asset management strategies, reduce downtime, and achieve better operational outcomes.

SAP Asset Intelligence Network, an Internet of Things (IoT) application, is seamlessly integrated into SAP BTP. It is a global equipment registry employing standardized definitions shared among various business partners, including manufacturers, original equipment manufacturers (OEMs), operators, and service providers. This collaborative sharing leads to the development of innovative business models that drive real operational excellence.

SAP Asset Intelligence Network offers a virtual platform for fostering collaboration on products and assets. The interconnected network of digital twins facilitates secure access, sharing, and governance of data on a global scale.

The following are some notable features.

- Facilitating secure collaboration within your business network

- Enabling collaboration among manufacturers, operators, and service providers across the network regarding asset information

- Streamlining collaborative efforts through a unified view

- Establishing a single network channel for the electronic transfer of technical assets and maintenance data

- Enhancing the reliability of data

- Reducing the need for frequent master data updates

Manufacturers contribute by releasing models with the desired content, encompassing manuals, drawings, installation instructions, repair guides, spare parts information, attributes, and counters. Suppliers play a role by providing spare parts data and other relevant content.

Asset data and associated information can be collaboratively exchanged through the SAP Asset Intelligence Network, referred to as "shared data," and synchronized with interconnected SAP ERP systems. Shared data updates are seamlessly integrated into the SAP ERP system upon equipment sharing or the publication of revised shared equipment models.

- Designing and generating equipment models

- Ensuring equipment synchronization between the cloud and back end

- Utilizing the Lookup feature

- Employing Smart Matcher for maintenance processes

- Initiating model requests

- Creating equipment requests

SAP Asset Intelligence Network offers a virtual platform for fostering collaboration on products and assets. The interconnected network of digital twins facilitates secure access, sharing, and governance of data on a global scale. Notable features include the following.

- Facilitating secure collaboration within your business network

- Enabling collaboration among manufacturers, operators, and service providers across the network regarding asset information

- Streamlining collaborative efforts through a unified view

- Establishing a single network channel for the electronic transfer of technical assets and maintenance data

- Enhancing the reliability of data

- Reducing the need for frequent master data updates

Manufacturers contribute by releasing models with the desired content, encompassing manuals, drawings, installation instructions, repair guides, spare parts information, attributes, and counters. Suppliers play a role by providing spare parts data and other relevant content.

Asset data and associated information can be collaboratively exchanged through the SAP Asset Intelligence Network and synchronized with interconnected SAP ERP systems. Shared data updates are seamlessly integrated into the SAP ERP system upon equipment sharing or the publication of revised shared equipment models.

The following object data can be shared.

- Equipment

- Functional Location

- Model

- Spare Part

- System

- Announcement

- Document

- Improvement Request

- Instruction

- Failure Mode (with Cause and Effect)

- Indicator

- Group

- Template

- Function

- Alert Type and Alert Type Group

- Maps and Geospatial data

- Notification

- Work order

8.3.1. Business Benefits

SAP Asset Intelligence Network provides several business benefits (see Figure 8-3) for manufacturers, asset owners, and service providers.

The following are some of the advantages to manufacturers.

- **Enhanced collaboration**: Manufacturers can collaborate more effectively with various stakeholders, such as asset owners and service providers, which improves communication and alignment.

- **Product visibility**: They can showcase their products more comprehensively by sharing detailed information like manuals, drawings, installation instructions, and repair guides, enhancing customers' understanding and confidence in their products.

- **Faster time-to-market**: Manufacturers can expedite the introduction of new products by streamlining the distribution of technical data, which accelerates the equipment's integration and deployment.

- **Market expansion**: By participating in a collaborative network, manufacturers can access new markets and customers, expanding their reach and potential revenue streams.

- **Data-driven insights**: The network enables manufacturers to gather insights from shared data to inform product improvements and innovation based on real-world usage and performance.

The following are some of the advantages to asset owners.

- **Centralized information**: Asset owners benefit from having a single, unified platform for managing and accessing information related to their assets, reducing the complexity of maintaining multiple data sources.

- **Optimized maintenance**: Access to up-to-date technical information, maintenance instructions, and spare parts details helps asset owners optimize their maintenance processes, reducing downtime and operational disruptions.

- **Improved decision-making**: Real-time access to asset data enables more informed decision-making, supporting strategies like predictive maintenance and lifecycle management.

- **Collaborative network**: Asset owners can interact directly with manufacturers and service providers on the platform, fostering collaboration and enhancing overall operational efficiency.

- **Data accuracy**: By utilizing standardized data shared on the network, asset owners can improve the accuracy and reliability of their asset information, reducing errors and minimizing risks.

The following are some of the advantages to service providers.

- **Efficient service delivery**: Service providers can access accurate and detailed information about assets, enabling them to offer better and more targeted maintenance and repair services.

- **Proactive maintenance**: With access to real-time data, service providers can transition from reactive to proactive maintenance strategies, addressing potential issues before they lead to costly breakdowns.

- **Collaboration**: Collaboration with manufacturers and asset owners allows service providers to better understand clients' needs and offer tailored solutions, fostering stronger customer relationships.

- **Reduced downtime**: Quick access to relevant technical information and spare parts details contributes to faster service delivery, minimizing asset downtime and improving overall operational continuity.

- **Business growth**: By participating in a collaborative ecosystem, service providers can expand their customer base, improve service quality, and ultimately grow their business.

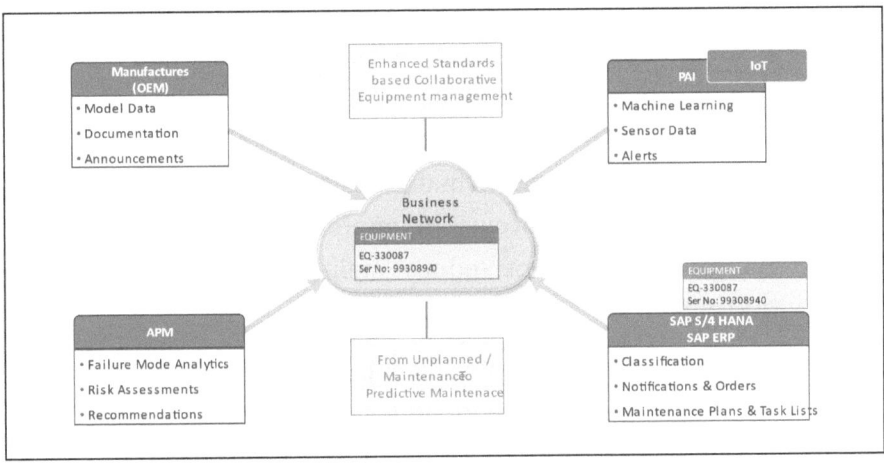

Figure 8-3. *SAP Business Network Asset Collaboration benefits*

In summary, the SAP Asset Intelligence Network facilitates enhanced collaboration, streamlined information sharing, improved maintenance practices, and greater business opportunities for manufacturers, asset owners, and service providers, leading to operational excellence and mutual growth.

8.4. SAP Predictive Asset Insights

SAP Predictive Asset Insights is a predictive maintenance and asset management solution. It is designed to help businesses proactively manage their assets and equipment by utilizing advanced analytics and predictive modeling techniques (see Figure 8-4).

SAP Predictive Asset Insights is built to address the following key aspects.

- **Predictive maintenance**: The solution uses data collected from various sources, such as sensors and operational systems, to analyze the performance and condition of assets in real time. Predictive analytics aims to forecast when maintenance is needed, allowing organizations to perform maintenance activities before equipment failures occur. This approach helps prevent unexpected downtime, reduces maintenance costs, and enhances operational efficiency.

- **Anomaly detection**: Using machine learning algorithms, the solution identifies anomalies or deviations from normal asset behavior. It can detect patterns that indicate potential issues or irregularities in asset performance, alerting maintenance teams to investigate and take necessary actions.

- **Data visualization and insights**: SAP Predictive Asset Insights provides visualizations and insights based on the analyzed data. These insights help maintenance teams and decision-makers understand asset health, usage trends, and potential risks more easily, enabling them to make informed decisions.

- **Integration**: The solution is typically integrated with various data sources, such as the IoT sensors, historical maintenance data, and other operational systems. This integration ensures that it has access to relevant and up-to-date information for accurate predictive modeling.

- **Cloud-based solutions**: SAP Predictive Asset Insights is a cloud-based solution that allows organizations to deploy and manage it more easily without significant on-premises infrastructure requirements.

SAP Predictive Asset Insights is an IoT application with an array of functions accessible via the SAP Fiori launchpad. The associated business advantages encompass the following.

- Mitigating costly disruptions through anticipating equipment malfunctions is achieved by processing extensive volumes of information technology (IT) and operational technology (OT) data using advanced machine learning algorithms

- Harnessing insights from sensor data enhances product quality, reliability, and overall customer satisfaction

- Effective management of intricate asset (equipment) structures

- Establishment of an asset network collaboration, promoting enhanced service and maintenance processes

SAP PAI offers the following functionalities.

- Sensor integration and alert management

- Utilization of machine learning techniques

- Analytical capabilities

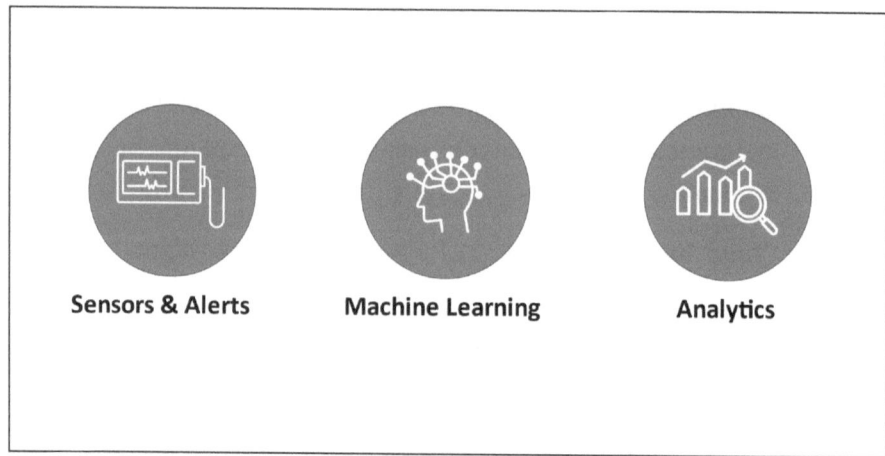

Figure 8-4. *SAP PAI capabilities*

8.5. SAP Asset Performance Management

SAP Asset Performance Management is a software solution provided by SAP that aims to help organizations effectively manage their asset strategies, optimize maintenance processes, and enhance overall asset performance. The following are key components and features typically associated with SAP Asset Performance Management.

- **Strategy formulation**: The solution assists in creating asset management strategies by considering factors such as asset criticality, risk assessment, and regulatory compliance. It helps organizations develop maintenance plans and strategies that align with business goals.

- **Performance monitoring**: It enables real-time monitoring of asset performance and health. The solution offers insights into asset conditions and performance metrics by integrating data from various sources, including sensors and operational systems.

- **Predictive analytics**: The solution incorporate spredictive analytics to anticipate potential asset failures and issues. Analyzing historical data and utilizing machine learning algorithms can forecast when assets might require maintenance or replacement.

- **Condition-based maintenance**: Based on real-time data and predictive insights, organizations can shift from traditional time-based maintenance to condition-based maintenance. This approach ensures maintenance is performed when necessary, reducing downtime and unnecessary maintenance activities.

- **Risk management**: The solution helps identify and assess risks associated with assets. By understanding potential risks, organizations can make informed decisions about maintenance priorities and resource allocation.

- **KPI tracking**: Key Performance Indicators (KPIs) are tracked to evaluate the effectiveness of asset management strategies and processes. This assists in measuring the impact of decisions on asset performance and overall business outcomes.

- **Integration**: SAP Asset Performance Management can be integrated with other enterprise systems, such as Enterprise Asset Management (EAM) or Enterprise Resource Planning (ERP) systems, to ensure seamless data flow and synchronization of asset-related information.

- **Data visualization:** The solution offers dashboards
  and visualizations that enable users to monitor asset
  performance, view historical trends, and make data-
  driven decisions.

Overall, SAP Asset Performance Management aims to empower
organizations to make proactive and informed decisions regarding their
asset management strategies, leading to improved reliability, reduced
operational costs, and enhanced operational efficiency. It is designed
to optimize asset performance, extend asset lifecycles, and align asset
management with broader business objectives.

SAP Asset Performance Management (see Figure 8-5) empowers
reliability engineers to evaluate asset health indicators and ongoing
performance, aiding in assessing maintenance strategy efficacy.

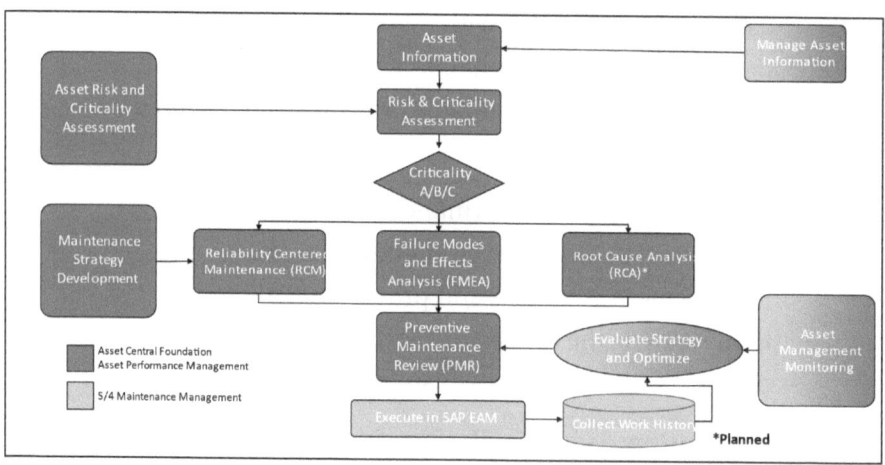

Figure 8-5. *SAP Asset Performance Management*

SAP APM encompasses the following functionalities.

- Identification of vital assets
- Generation of risk scores

- Development of assessments

- Selection of the optimal analytical approach (RCM, FMEA, PM Review)

- Provision of maintenance recommendations rooted in RCM and FMEA insights

Criticality is contingent upon the level of risk. Risk is conventionally computed by multiplying the probability of failure by the consequence of said failure. Consequently, heightened risk corresponds to heightened equipment criticality. The assessment of criticality relies on the definition of predetermined thresholds. This process accommodates the definition of technical and financial risks, which can be specified at the dimension level within an assessment template.

Normalized risk is a percentage figure that represents risk. Its purpose is to enable reliability engineers to compare the risk associated with an "air compressor" against that of an "electrical bulb." The intention is to equalize the evaluation for both objects.

The following describes risk type scores across the risk lifecycle.

- **Current/existing/initial risk**: This constitutes the risk assumed during the installation of new equipment or when conducting an initial risk assessment.

- **Mitigated risk**: This quantifies the amount of risk that has been reduced after executing actions like work orders, redesigns, replacements, or training.

- **Unmitigated risk**: This signifies the residual risk between the other two stages.

Figure 8-6 illustrates the steps to implement SAP APM.

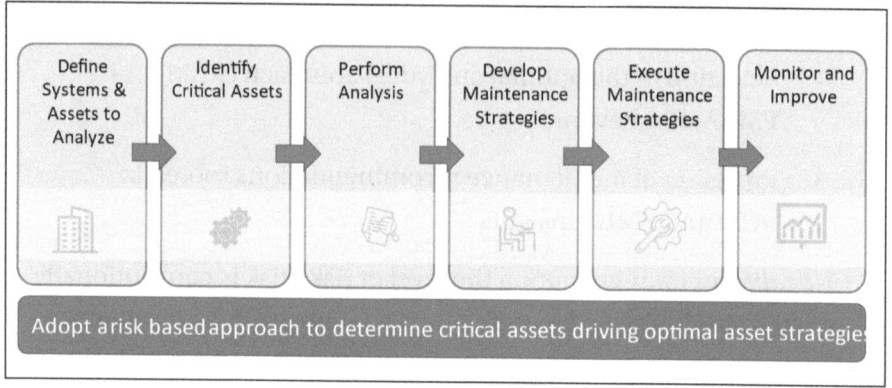

Figure 8-6. *Steps to implement SAP APM*

1. Specify systems and assets for analysis. Asset Central Foundation facilitates the structuring of asset data using templates aligned with established ISO standards such as ISO 14224. This encompasses models, equipment, locations, systems, groups, spare parts, documents, and failure modes, all integrated within Asset Central Foundation.

2. Determine critical assets. Through risk and criticality assessment, assets are arranged in order of risk and criticality, serving as a foundation for subsequent analysis. This entails constructing a matrix that juxtaposes the consequence of failure against the probability of failure. Following the execution of assessments (e.g., for specific equipment), the risk score and risk type score difference can be accessed within the Assessment tab's Highlights screen.

3. Conduct an analysis. Employ well-established methodologies such as reliability-centered maintenance (RCM), failure modes and effects analysis (FMEA), risk-based inspection (RBI), and root cause analysis (RCA) to discern the most suitable maintenance approaches for your assets.

4. Formulate recommended actions. The outcome of RCM/FMEA yields suggestions for preventive and/or corrective tasks. The Preventive Maintenance Review (PMR) application oversees these recommendations, facilitating the creation of published instructions.

5. Generate recommendations as the conclusive phase of the RCM or FMEA assessment based on the following options.

    • Existing Instructions

    • Imported Preventive Maintenance Tasks (Task Lists "filtered" by Technical Object Maintenance Plans)

    • Placeholder Instructions (utilize existing or generate new ones)

    • Imported Tasks (Task Lists)

6. Construct or associate instructions with the recommendations.

7. Upload them to S/4HANA using a .csv file.

8. Execute maintenance strategies. Enact the instructions within the maintenance management system.

9. Monitor and enhance. Continuously oversee the performance and efficacy of the maintenance strategies.

8.6. SAP Mobile Asset Manager

SAP Mobile Asset Manager or Asset Manager is a mobile application developed by SAP. It is designed to enable organizations to effectively manage and maintain their assets using mobile devices. This application is typically integrated with SAP's broader asset management solutions, such as EAM or PM modules, to extend asset management capabilities to field technicians and mobile users.

The following are key features and functions associated with SAP Mobile Asset Manager.

- **Mobile asset management**: The application allows field technicians and maintenance personnel to access critical asset information and perform various asset management tasks directly from their mobile devices. This includes viewing equipment details, recording maintenance activities, and updating asset status.

- **Work management**: Field technicians can receive and manage work orders on mobile devices. This enables them to view work order details, access relevant documentation, update work order status, and report the completion of tasks in real time.

- **Asset inspections**: The application supports asset inspections, allowing users to perform routine inspections, capture inspection data, and report any issues or anomalies directly from the field.

- **Offline capabilities**: SAP Mobile Asset Manager offers offline functionality, enabling users to continue working even in areas with limited or no internet connectivity. Data captured offline can be synchronized with the central system once a connection is restored.

- **Barcode and QR code integration**: The application could integrate with barcode or QR code scanning capabilities, making it easier for users to quickly identify and access asset details by scanning asset labels.

- **User-friendly interface**: The user interface is intuitive and user-friendly, accommodating field workers with varying technical expertise.

Integration with SAP EAM/PM: SAP Mobile Asset Manager is designed to seamlessly integrate with SAP's EAM or PM modules, ensuring that data captured through the mobile app is synchronized with the central asset management system.

SAP Asset Manager emerges as our cutting-edge mobile application, centered around assets, effectively leveraging the digital core. It seamlessly integrates with SAP EAM, SAP S/4HANA Asset Management (On-Premise), and SAP IAM solutions.

The architecture of SAP Asset Manager relies on the Mobile Development Kit, coupled with SAP BTP SDK for iOS and Android. These frameworks follow the Fiori for iOS and Fiori for Android guidelines, culminating in a streamlined user experience.

SAP Asset Manager is a metadata-driven application that generates code supporting deployment across various platforms. This deployment is facilitated through middleware utilizing SAP BTP, managing data interactions between the back end and mobile devices while offering offline capabilities.

Why might a customer opt for SAP Asset Manager over SAP Work Manager? SAP Asset Manager facilitates digital transformation by harnessing the advanced capabilities of SAP BTP, coupled with the user-centric design language embodied by Fiori for iOS and Fiori for Android.

The SAP Asset Manager application efficiently oversees tasks encompassing work orders, notifications, condition monitoring, material consumption, time management, and failure analysis.

SAP Asset Manager provides the following benefits.

- Secure the capture and retrieval of precise, up-to-date information, whether online or offline, to effectively address rising market demands, globalization, and pressures related to regulations, social aspects, and the environment.

- Enhance the management of intricate assets and dependencies on external parties.

- Safeguard and expand your organization's knowledge repository while mitigating employee turnover's impact.

8.6.1. SAP Mobile Add-On

The SAP Mobile Add-On is the back-end system responsible for processing all data related to the SAP Asset Manager application.

This "mobile add-on" refers to supplementary components that provide additional functionalities beyond the primary SAP product. These components are incorporated later by external, independent entities or by SAP itself, tailoring them to customers' specific requirements.

Positioned above the core, this mobile add-on interfaces with the same dictionary or repository objects while executing the necessary operations.

From a user perspective, there is no noticeable distinction on your mobile device when utilizing a system integrated with a mobile add-on versus one that is not. Nonetheless, understanding the terminology is valuable.

The fundamental capabilities of Standard Asset Management encompass the following.

- Installation and dismantling of equipment

- Management of work order and notification lifecycles

- Clock In and Clock Out functionalities for multiple users collaborating in the same order

- Documenting measurement points

- Adding and issuing components

- Support for object lists

- Recording Cross-Application Time Sheet (CATS) time reports

- Confirmations for preventive maintenance (PM)

- Implementation of classifications

- Incorporation of business partner and warranty details for equipment

8.6.2. SAP Mobile Development Kit (MDK)

A constituent of SAP BTP Mobile Services, the mobile development kit empowers customers to create fresh native mobile applications and tailor specific intricate mobile applications from SAP. This augmentation of our mobile services introduces a metadata-driven approach to mobile app development.

SAP Asset Manager takes the lead as the inaugural app to leverage the SAP Enterprise App Modeler tool. The initial release of the SEAM (SAP Enterprise App Modeler) is primarily geared toward personalizing SAP Asset Manager. At the same time, subsequent iterations are poised to encompass support for novel complex mobile applications. Additionally, they aim to enable the swift creation of customized applications.

In this development mode, developers operate at a significantly higher level of abstraction than crafting applications using Java or Swift. The intricacies of low-level system details are alleviated. This framework facilitates a broader spectrum of resources for application development, offering an abstraction layer conducive to cross-platform development.

8.7. SAP Asset Central Foundation

SAP Asset Central Foundation is a component of SAP's suite of asset management solutions that aims to provide a comprehensive and unified platform for managing and maintaining a wide range of assets within an organization (see Figure 8-7). It is designed to support various industries and sectors, enabling organizations to effectively manage their assets throughout their lifecycle.

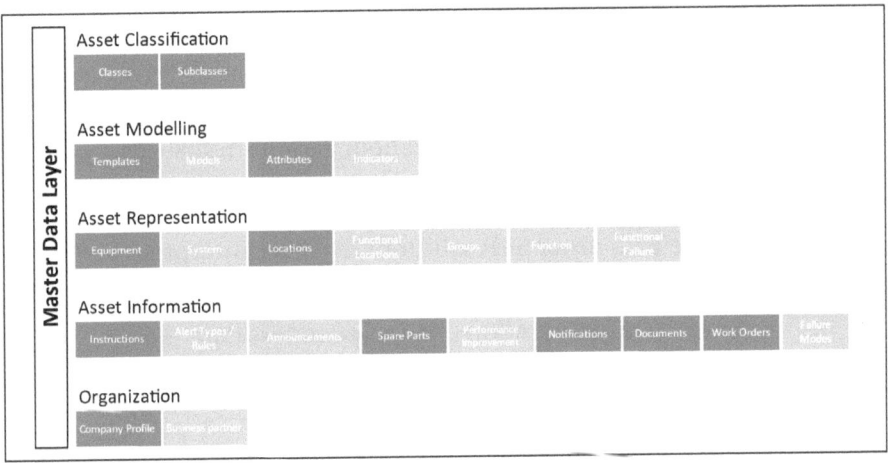

Figure 8-7. *SAP Asset Central master data*

The following are key features and aspects typically associated with SAP Asset Central Foundation.

- **Unified asset management**: SAP Asset Central Foundation offers a centralized platform to manage diverse types of assets, such as equipment, machinery, vehicles, and infrastructure, regardless of industry or location.

- **Data modeling**: The solution facilitates the structuring of asset-related data using templates based on standardized schemas and models. This ensures consistency in asset data representation across the organization.

- **Metadata driven**: Asset Central Foundation is often driven by metadata, allowing organizations to define attributes, relationships, and behaviors of assets. This enables flexible adaptation to specific asset management requirements.

- **ISO standards**: The solution may support integration with ISO standards such as ISO14224, which provides guidelines for collecting and exchanging reliability and maintenance data for equipment in various industries.

- **Integration**: Asset Central Foundation is designed to integrate with other SAP solutions, such as Enterprise Asset Management (EAM), PM, and other asset-related modules. This integration ensures seamless data flow between asset management processes and other business functions.

- **User-friendly interface**: Asset Central Foundation offers an intuitive user interface that allows users to easily navigate and interact with asset data. This contributes to improved user adoption and efficiency.

- **Reporting and analytics**: The solution may provide tools for generating reports and performing analytics on asset data. This helps organizations make informed decisions and identify areas for improvement in asset management.

- **Lifecycle management**: Asset Central Foundation supports the management of assets throughout their entire lifecycle, from acquisition and installation to operation, maintenance, and eventual disposal.

8.8. SAP IAM Integration

SAP IAM Integration refers to integrating SAP's IAM solutions with other systems, applications, and data sources within an organization's technology landscape. SAP IAM solutions are designed to optimize asset performance, maintenance processes, and operational efficiency through the use of advanced technologies such as the IoT, predictive analytics, AI, and machine learning. The SAP system landscape is illustrated in Figure 8-8.

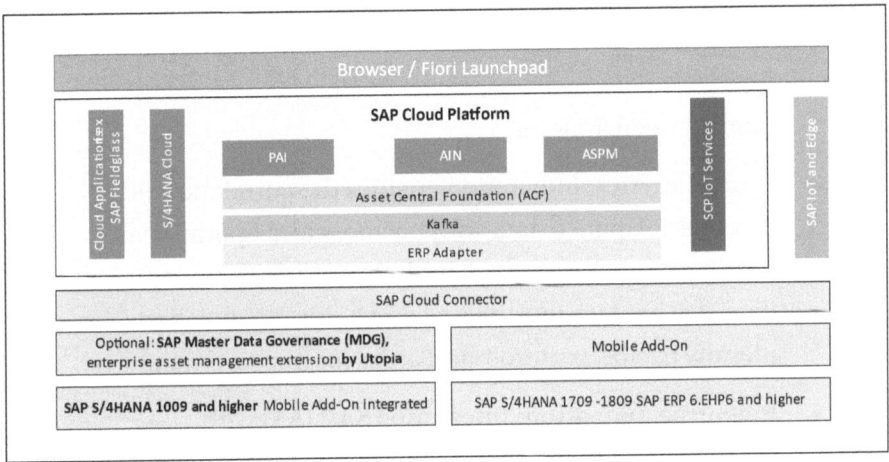

Figure 8-8. *SAP system landscape*

The following are key aspects and benefits of SAP IAM integration.

- **Data integration**: Integration involves connecting SAP IAM solutions with various data sources, including IoT sensors, operational systems, historical maintenance data, and more. This enables comprehensive and real-time insights into asset health, usage patterns, and performance metrics.

- **Predictive maintenance**: By integrating IoT data and predictive analytics, SAP IAM solutions can anticipate potential equipment failures and maintenance needs. This proactive approach helps organizations perform maintenance tasks optimally, reducing downtime and operational disruptions.

- **Work order management**: Integrating SAP IAM with other systems, such as Enterprise Asset Management (EAM), allows organizations to seamlessly manage work orders, notifications, and maintenance tasks. This integration streamlines communication between field technicians, maintenance personnel, and asset management systems.

- **Data analytics**: Integration enables the correlation of asset performance data with business and operational data from other systems. This holistic view supports data-driven decision-making and helps organizations identify trends, patterns, and areas for improvement.

- **Resource allocation**: Integrating SAP IAM with resource planning systems can optimize the allocation of personnel, equipment, and materials for maintenance tasks, leading to better resource utilization and cost savings.

- **Collaboration**: SAP IAM integration facilitates collaboration across different departments, such as maintenance, operations, finance, and procurement. It ensures that all relevant stakeholders have access to up-to-date asset information.

- **Mobile access**: Integration can extend SAP IAM capabilities to mobile devices, allowing field technicians to access asset information, perform inspections, and update maintenance records in real time.

- **Regulatory compliance**: Integrating SAP IAM with compliance and regulatory systems helps ensure that asset management practices adhere to industry standards and regulations.

- **Asset lifecycle management**: Integration supports seamless data flow throughout the asset lifecycle, from acquisition and installation to operation, maintenance, and eventual disposal.

8.9. Summary

This chapter offered a comprehensive overview of SAP's advanced asset management solutions, encompassing key chapters such as SAP IAM, SAP Business Network with Asset Collaboration, SAP Asset Predictive Insights, SAP Asset Performance Management, SAP Mobile Asset Management, and SAP IAM integration. The chapter began by highlighting the significance of SAP IAM, showcasing how it harnesses IoT and AI technologies to optimize asset utilization and maintenance. It then delved into SAP Business Network with Asset Collaboration, illustrating how collaborative platforms facilitate streamlined stakeholder communication for improved asset productivity. The discussion extends to SAP Asset Predictive Insights, emphasizing its role in utilizing predictive analytics to forecast maintenance needs and mitigate downtime.

Furthermore, the chapter explored SAP Asset Performance Management, elucidating its importance in continuously monitoring, analyzing, and optimizing asset performance throughout its lifecycle. The inclusion of SAP Mobile Asset Management underscores the flexibility and efficiency offered by mobile applications in asset management processes. The chapter concluded with insights into SAP IAM Integration, demonstrating how these solutions seamlessly integrate with existing systems to create a holistic asset management framework. Overall, this chapter provided a comprehensive look at SAP's innovative approaches to modern asset management.

Index

A

B